99-PUM-2Ta

THE AGRONOMIC AND ENVIRONMENTAL AVAILABILITY OF BIOSOLIDS-P (PHASE II)

by:

George A. O'Connor
University of Florida

Herschel A. Elliott
Pennsylvania State University

2006

WERF

The Water Environment Research Foundation, a not-for-profit organization, funds and manages water quality research for its subscribers through a diverse public-private partnership between municipal utilities, corporations, academia, industry, and the federal government. WERF subscribers include municipal and regional water and wastewater utilities, industrial corporations, environmental engineering firms, and others that share a commitment to cost-effective water quality solutions. WERF is dedicated to advancing science and technology addressing water quality issues as they impact water resources, the atmosphere, the lands, and quality of life.

For more information, contact:
Water Environment Research Foundation
635 Slaters Lane, Suite 300
Alexandria, VA 22314-1177
Tel: (703) 684-2470
Fax: (703) 299-0742
www.werf.org
werf@werf.org

This report was co-published by the following organization. For nonsubscriber sales information, contact:

IWA Publishing
Alliance House, 12 Caxton Street
London SW1H 0QS, United Kingdom
Tel: +44 (0) 20 7654 5500
Fax: +44 (0) 20 7654 5555
www.iwapublishing.com
publications@iwap.co.uk

Library of Congress Catalog Card Number: 2005930913
IWAP ISBN: 978-1-84339-782-3

University of Florida, Pennsylvania State University

ACKNOWLEDGMENTS

The project team thanks the Water Environment Research Foundation and its subscribers for supporting this project. Members of the project subcommittee provided valuable input throughout the project and reviewed interim and final reports. WERF staff provided technical and, especially, moral support throughout the project. Special thanks, also, to Robin Brandt, Scott Brinton, and Olawale Oladeji for their assistance in conducting experiments key to the project's success, and to Maria Silveira for her help in preparing the report.

Report Preparation

Principal Investigators:
George A. O'Connor, Ph.D.
University of Florida

Herschel A. Elliott, Ph.D., P.E.
Pennsylvania State University

Project Subcommittee

David S. Taylor, Chair
Madison Metropolitan Sewerage District

Michael Jawson (Research Council Liaison)
USDA, ARS

John T. Gilmour, Ph.D.
University of Arkansas

John Walker, Ph.D.
U.S. Environmental Protection Agency

David R. Zenz, Ph.D., P.E.
CTE Engineers

Diane Gilbert
City of Los Angeles Bureau of Sanitation

Water Environment Research Foundation Staff

Director of Research: Daniel M. Woltering, Ph.D.
Program Director: Jami Montgomery

ABSTRACT AND BENEFITS

Abstract:

Concerns about water quality degradation resulting from land application of manures and biosolids have elicited a wide array of national and state regulations. The regulations focus on phosphorous (P) as the limiting nutrient to eutrophication of most freshwater supplies and threaten to severely impact biosolids recycling. Critical data about biosolids-P phytoavailability and runoff potential, especially field data, needed to validate the regulations are scarce. The goal of this research was to supply such data. Conventionally treated biosolids exhibit both less agronomic and environmental P lability than manures and fertilizers. Biosolids processed for biological P removal, however, mimic fertilizer-P. These biosolids require special management to reduce environmental impact, such as soil incorporation and co-application with water treatment residuals.

Benefits:

- Quantifies the relative phytoavailability of biosolids-P in terms of biosolids and soil characteristics
- Shows that water-extractable P is a good predictor of biosolids agronomic and environmental lability
- Supports inclusion of P-source coefficients in P indexes to access P environmental impact and to distinguish between various P-sources
- Proposes principles to sustain land application as a biosolids recycling strategy even under P-based nutrient management

Keywords: Nutrient management, P index, water quality, sustainability

TABLE OF CONTENTS

LIST OF TABLES

LIST OF FIGURES

LIST OF ACRONYMS

ANOVA	analysis of variance
Al-WTR	aluminum-based water treatment residual
BAP	biologically available phosphorus
BPR	biological phosphorus removal
DAP	diammonium phosphate fertilizer
DM	dairy manure
DMAD	dewatered, mesophilic, anaerobically digested
DPS	degree of phosphorus saturation
ICP	inductively coupled plasma atomic emission spectroscopy
ISP	iron-strip phosphorus
MWRDGC	Metropolitan Water Reclamation District of Greater Chicago
NH_4NO_3	ammonium nitrate fertilizer
PP	particulate phosphorus
PSC	phosphorus source coefficient
RPP	relative phosphorus phytoavailability
SFWMD	South Florida Water Management District
SM	swine manure
SRP	soluble reactive phosphorus
STP	soil test P
TDP	total dissolved phosphorus
TP	total phosphorus
TS	total solids
TSP	triple superphosphate (fertilizer-P)
U.S. EPA	U.S. Environmental Protection Agency
WTR	water treatment residual
WWTP	wastewater treatment plant
X_T	total element (X)

EXECUTIVE SUMMARY

Concerns about water quality degradation associated with land application of manures and biosolids have elicited a wide array of national and state regulations. The regulations focus on P as the limiting nutrient to eutrophication of most freshwater supplies. Mandated P-based nutrient management could significantly constrain biosolids recycling programs. The goal of this project was to continue efforts begun in Phase I (O'Connor et al., 2002) to characterize the form, solubility, and bioavailability of P in biosolids and to expand into consideration of P transport issues. The intent was to generate data that foster rational nutrient management policies for land-based recycling.

Phase I work showed that inorganic forms of P dominate in manures and in both conventional and "enhanced" [biological P removal (BPR), heat-dried] biosolids. Organic forms play a minor role in determining the agronomic and environmental lability of P. The short-term phytoavailability of P in conventionally prepared biosolids was about 50% that of fertilizer-P (triple superphosphate, or TSP), consistent with the 50% "relative effectiveness" factor recommended by U.S. EPA (1995). Biosolids prepared via BPR, however, mimicked fertilizer-P in lability. Heat-dried materials containing high levels of total Fe + Al were much less ($\leq$ 10%) labile. Similarly, co-applying biosolids with Fe- or Al-based water treatment residuals (WTRs) dramatically reduced biosolids-P lability.

Phase II work sought to validate the laboratory and greenhouse data from Phase I with field data. In addition, particular attention was given to relating residuals source characteristics with the potential to deliver P to water bodies. Principles were proposed to sustain land application as a biosolids recycling strategy under P-based nutrient management.

Limited field data are available to validate biosolids-P phytoavailability, especially in the long-term. Data from a one-year study in the United Kingdom (Smith et al., 2003) appear to confirm that BPR biosolids can result in relative P phytoavailability (RPP) values equal to, or exceeding, fertilizer-P. Conventional (dewatered, mesophilic anaerobically digested – DMAD, and Fe-dosed DMAD) materials had RPP values of 53–59%. The UK study was conducted on soil containing abundant native P and complicated data interpretation.

Data were also collected from an ongoing field study funded by the South Florida Water Management District (SFWMD). The study involved the same crop (bahiagrass; *Paspalum notatum* Flugge) and the P-deficient soil (Immokalee fine sand) used in Phase I greenhouse studies. The resulting RPP values are highlighted in Table ES-1, which also includes greenhouse results from Phase I. The RPP values of the Boca Raton and Pompano biosolids and the poultry waste were assigned relative phytoavailability categories developed in Phase I work. The category assignment paralleled differences in water-extractable P values of the materials. This supports the hypothesis that material water-extractable P is a good indicator of material-P phytoavailability on initially P-deficient soils. The limited field data available appear to validate the greenhouse data of Phase I. Biosolids total P, neutral ammonium citrate-P, or common soil test P failed to indicate RPP values well. A laboratory study of 35 biosolids (Elliott et al., 2005b)

Table ES-1. Relative P Phytoavailability (RPP), Category of Relative Phytoavailability, and Water-Extractable P of P-Sources Used in Greenhouse and SFWMD Field (Bolded) Studies.

Source	RPP (%)	Category of Relative Phytoavailability	Water-Extractable P ($g\ kg^{-1}$)
TSP	100	High	21
Largo cake	74	High	8.8
Wisconsin slurry	104	High	6.2
Boca Raton biosolids	**85**	**High**	**5.2**
Poultry waste	**55**	**Medium**	**3.5**
Tarpon Springs cake	31	Medium	3.4
Largo pellets	48	Medium	2.6
Untreated aerobic cake	54	Medium	1.7
Pompano biosolids	**59**	**Medium**	**1.4**
Baltimore cake	34	Medium	0.86
Philadelphia cake	37	Medium	0.57
Untreated anaerobic cake	26	Medium	0.29
Baltimore pellets	9	Low	0.17
Philadelphia compost	7	Low	0. 05

concluded that the neutral ammonium citrate extraction method, used to measure plant available P for mineral fertilizers, effectively measured biosolids total P and thus was inappropriate for quantifying biosolids phytoavailable P.

A recent greenhouse study (Ebeling et al., 2003) using wheat as the test crop also found the Wisconsin slurry (a BPR) to result in RPP values 130–170%, depending on the soil used. Differences in biosolids RPP values are best expressed on P-responsive soils (initially P-deficient and/or having low P sorbing capacity). Many soils used for recycling programs have neither of these characteristics (are not P-responsive) owing to historical fertilization and waste management practices. In greenhouse studies with such soils (Sarkar and O'Connor, 2005), differences in biosolids RPP (and water-extractable P) values are largely masked by the soil, so most P-sources (including fertilizer-P) appear to have similar agronomic effectiveness. BPR materials, however, consistently yield large RPP values in the first growing season following application when BPR materials are applied at high rates mimicking N-based biosolids applications. The reason for this surprising result is not clear, but may be caused by the immediate solubility of BPR-P compared to even fertilizer-P, which requires time for complete dissolution and plant uptake. There are no long-term data currently available to test the hypothesis.

The "low" relative phytoavailability category (RPP values ≤ 10%) of biosolids could not be unequivocally validated with field data. The low RPP values measured in Phase I work suggest that pelletized, high total Fe + Al materials could be applied at N-based rates to achieve agronomic benefit equal to fertilizer-P. The long-term availability of P in these materials, however, has not been studied. On most soils used for recycling biosolids, agronomic effectiveness of biosolids-P from material with low and medium RPP values is probably of minor importance. Of much greater interest is the environmental lability of biosolids-P, and this was the subject of our Phase II work.

Indoor and field rainfall simulation experiments showed that there are significant differences in runoff P concentrations from surface applied manures and biosolids. When applied at a common total P rate, P losses generally decreased in the order: poultry and swine manures > dairy manure and BPR biosolids > non-BPR biosolids. The P concentrations in runoff from biosolids-amended soils depend on the types of wastewater and solids processing methods used to generate the biosolids. Some biosolids types (e.g., BPR products) resulted in higher runoff P concentrations than dairy manure applied at the same plant available N rate. Other biosolids, because of additions of Al and/or Fe during wastewater treatment, or through solids processes like heat drying, produced runoff P losses indistinguishable from unamended soil treatments. The data provide strong evidence that the solubility of the P in the organic amendment, as measured by its water-extractable P, exerts a major influence on the potential for offsite migration of P at land application sites.

P runoff losses were reduced when WTRs were co-applied with various P sources (fertilizer, manures, and biosolids). Adequate contact time between the P source and WTR is necessary to effect significant reduction in P losses. For a BPR material, runoff losses (total dissolved P and total P) were high following surface spreading, but were not different ($\alpha = 0.05$) from unamended soil when the biosolids were incorporated before rainfall application.

Runoff P losses from surface applied biosolids and manures increase with application rate for a single P source. However, when multiple P sources are considered collectively, a positive correlation between runoff P and the total P application rate was not found (Figure ES-1) because of the widely varying solubility of the P in different biosolids. However, when the application rate is adjusted to account for the solubility of the P source using a weighting coefficient (P source coefficients; PSCs), a strong correlation with runoff P losses was found (Figure ES-2). Inclusion of PSCs in P site indices is particularly important for biosolids and treated manures where the susceptibility of P to leaching and runoff is highly variable. Unless the variable and typically lower solubility of P in biosolids is taken into account, the potential for P loss from biosolids application fields will likely be overestimated by site index assessment tools.

Despite the progress made in Phase I and II work, several challenges remain.

1. During the last decade, many wastewater treatment plants (WWTPs) have been designed to remove nutrients biologically (Tchobanoglous et al., 2003). Because of ease of implementation for existing conventional activated wastewater treatment facilities, BPR is rapidly emerging as an important technology to meet effluent P discharge permits (Daigger, 2003). However, BPR-generated biosolids have exceptionally high P solubility and loss potential when surface applied. Treatment strategies and management guidelines are needed to ensure that land-based recycling continues as an option for treatment plants employing BPR.

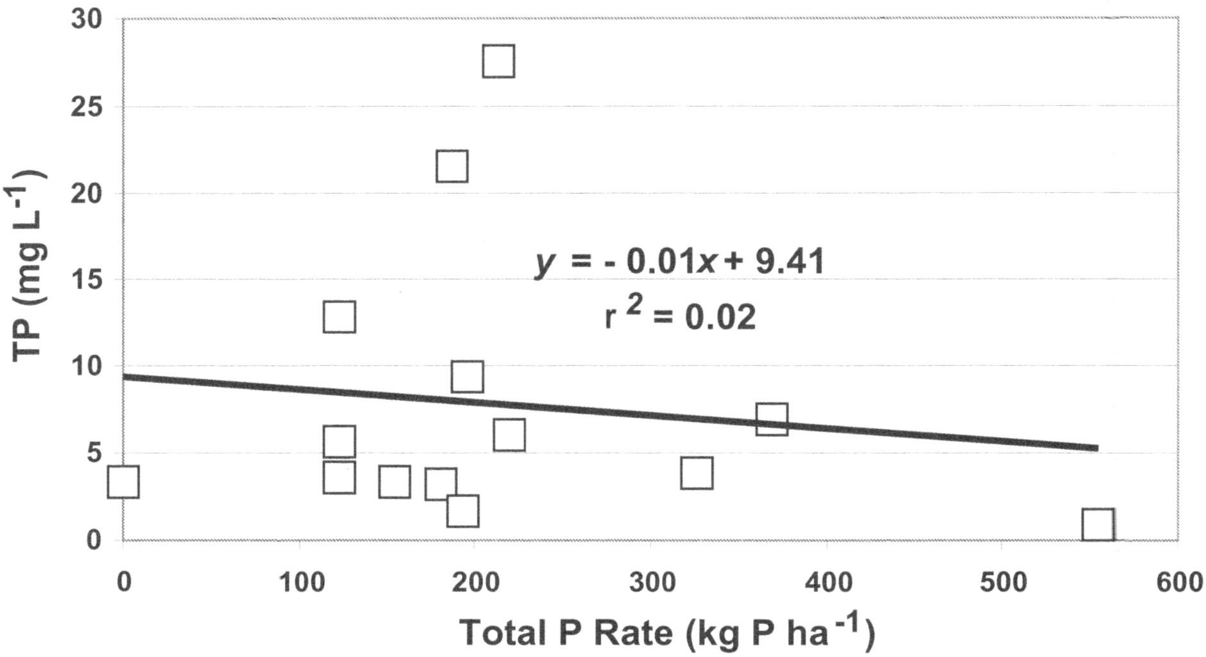

Figure ES-1. Total Runoff P Concentration for Biosolids Applied at a Common Plant-Available N Rate.

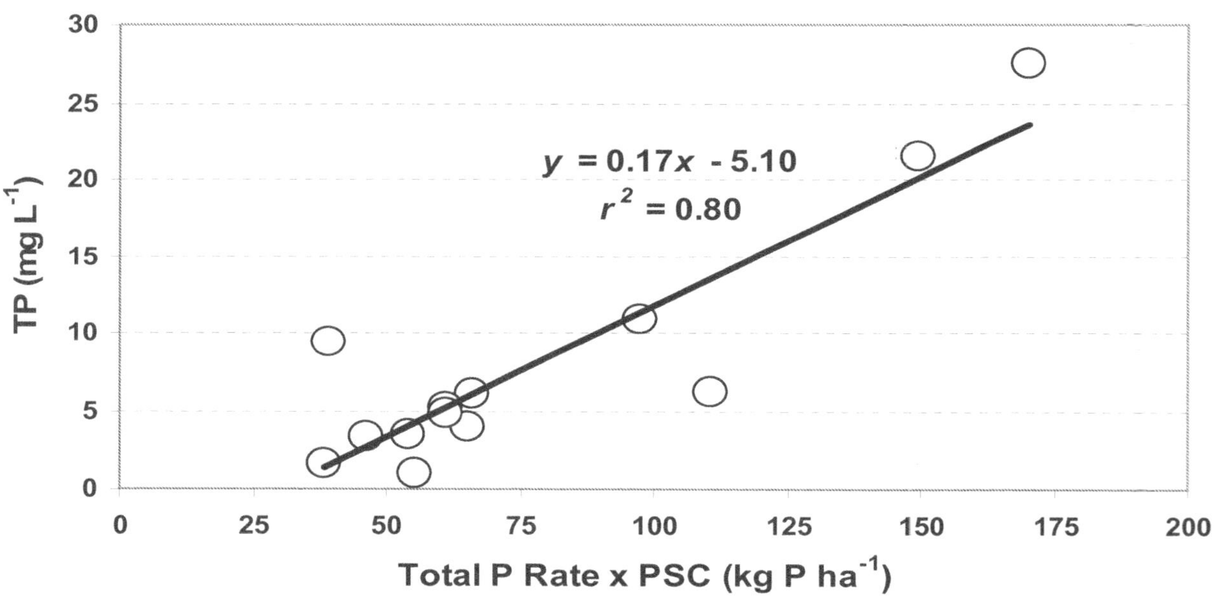

Figure ES-2. Relationship Between Total Runoff P Concentration and the P Source Coefficient-Adjusted P-Source Application Rate.

2. Research to date has consistently shown that biosolids high in Al and/or Fe have relatively low potential for offsite P migration. The long-term stability of the high metal concentration effect, however, has not been verified. Of special concern is the fate of Fe-bound P compounds under anaerobic conditions, such as might exist if eroded biosolids-amended soil particles are transported to aquatic sediments.

3. Perhaps the greatest challenge nationally is developing consistent and scientifically sound regulatory strategies for biosolids under P-based nutrient management policies. Most states have P indices of varying types and complexity, the development of which has, and continues, to outstrip validating science. Biosolids professionals must provide input and conduct field research addressing the behavior and fate of P in land-based biosolids recycling.

CHAPTER 1.0

INTRODUCTION

The United States produces more than six million metric tons of biosolids annually (Epstein, 2003). This represents a substantial resource beneficially usable for crop production (nutrient source) and soil conditioning (organic matter source). Land application is the most common beneficial utilization method for biosolids. In land-based recycling, biosolids are typically applied at rates designed to satisfy crop nitrogen (N) requirements and avoid excessive nitrate leaching to groundwater. Because of the mismatch between biosolids nutrient [N and phosphorus (P)] contents and crop fertility requirements, P is normally supplied in excess of that needed for optimum crop growth. The excess P accumulates in the soil or is lost to the environment. Although the accumulated P is not an agronomic problem, loss of P to aquatic systems is a concern because P is the nutrient limiting eutrophication in most freshwater systems (Carpenter et al., 1998; Sharpley et al., 2003).

A recent survey by Shober and Sims (2003) found that 24 states have existing regulations that can limit biosolids application rates based on existing P levels in the soil. These states typically have established numerical environmental thresholds of soil test P levels beyond which biosolids application is prohibited. However, the current consensus among states is that a comprehensive approach using P site indices is a more effective way to identify agricultural fields vulnerable to P loss. These P site indices evaluate the relative risk to P loss based on site characteristics influencing P transport, the type of P source applied, and various crop and management practices (Lemunyon and Gilbert, 1993; Leytem et al., 2004). Forty-seven states have adopted the P index approach. Although efforts so far have targeted manures and commercial fertilizers, P-based management will likely apply to *all* land-applied residuals including biosolids. Relatively few states have explicitly addressed biosolids in P indices, despite the prevalence of land application for biosolids recycling. This is unfortunate because the solubility, bioavailability, and transport potential of P depend on whether the P originates from fertilizer, manure, compost, or biosolids (Withers et al., 2001; Leytem et al., 2004; Brandt et al., 2004). Coale et al. (2002) predicted that P indices will be deployed as regulation long before being objectively validated. Quantifying P losses from application of biosolids to agricultural land is critical to promote scientifically sound P management policies.

Although the details vary, most states have established interpretive categories that define nutrient management practices based on a site's P index score. One management category dictates that the soil application rates of organic by-products are restricted to meeting the P requirement of the crop (i.e., the P-based rate). Under this condition, accurate nutrient management planning requires accounting for the differences in phytoavailability of various biosolids. The U.S. EPA Process Design Manual for biosolids (U.S. EPA, 1995) acknowledges the variability in phytoavailable P in different materials and recommends using an average phytoavailability (relative effectiveness) factor of 50%. Canadian regulators (Ontario Ministry of Environment and Energy and Ontario Ministry of Agriculture, Food, and Rural Affairs, 1996) recommend a value of 40%. Both relative effectiveness factors tacitly acknowledge that not all the P in biosolids is phytoavailable. Indeed, biosolids-P phytoavailability is expected to depend on the residuals treatment process(es) used because they affect the form and solubility of P in biosolids (Pastene, 1981; Soon and Bates, 1982; Smith et al., 2002). Given the interest in using various metal salts to treat lagooned manures or poultry bedding (Shreve et al., 1996), different P phytoavailabilities should be expected in manures as well.

Residuals treatment processes continue to evolve, as do management practices such as combining different residuals to produce final products with more desirable environmental and agronomic properties. Examples include biological (microbial) P removal (BPR) treatment to reduce effluent P concentrations (Loosdrecht, 1998) and combining biosolids and water treatment residuals (WTRs) to reduce P solubility in the resulting mixture (Elliott et al., 2002). The effect(s) of these (and other) processes and management practices on P form, solubility, and phytoavailability can be inferred, but must be quantified to appropriately address P phytoavailability and environmental issues.

Phase I of this project (O'Connor et al., 2002) addressed several of the important issues discussed above, including characterizing P form and solubility in biosolids and manures and in soils amended with both residuals, determining the phytoavailability of various P sources, deriving procedures to characterize available P in residuals, and determining the effect of co-applying WTRs and biosolids on P lability. Significant findings from Phase I work were

- Most ($\geq$ 80%) total P in biosolids and manures is inorganic. Organic forms in residuals, and especially residuals-amended soils, are generally of minor importance.
- The short-term phytoavailability of biosolids-P is typically 40–50% of fertilizer-P. Notable exceptions are heat-dried and high total Fe and Al materials, which have much lower phytoavailability, and biological P removal (BPR) biosolids, which have phytoavailability similar to fertilizer-P.
- The total P concentration, traditional soil tests, and fertilizer-P tests are unreliable measures of labile P in biosolids. Water-extractable P is well correlated to labile P, and is typically $\leq$5% total P for many biosolids.
- WTRs (Al- or Fe-based) can effectively control P leaching in sandy soils amended with soluble P sources (fertilizer, manure, or BPR biosolids). Leaching losses are minimal for most biosolids.

A major conclusion from Phase I work was that results needed to be validated, especially in the field, and that inferences drawn from the suite of residuals and soils examined needed to be expanded to greater national significance. Accordingly, the two guiding objectives of Phase II research are

1. Validate laboratory and greenhouse results from Phase I by collecting samples and raw data from completed and ongoing field studies.

2. Determine and compare the P-runoff potential of biosolids-, manure-, and fertilizer-P-amended soils, alone and in the presence of WTRs using rainfall simulations.

Particular attention has been given to relating nutrient source characteristics with the potential to deliver P to water bodies. Principles are proposed to sustain land application as a biosolids recycling strategy under P-based nutrient management.

CHAPTER 2.0

METHODS

2.1 Field and Greenhouse Studies

2.1.1 SFWMD Field Project

The project team attempted to utilize data from completed and ongoing studies to validate the phytoavailability results of Phase I (O'Connor et al., 2002). No completed field studies were identified that provided unambiguous P phytoavailability data useful to the project, but one study (Smith et al., 2003) appears to provide useful data. Only one ongoing (now completed) field study was identified that partially met project needs. The project was funded by the SFWMD and sought to document environmental problems associated with land-applied biosolids and chicken manure, and evaluate the efficacy of an aluminum-based (Al)-WTR as a best management practice to minimize P losses from Florida cattle pastures.

The South Florida Water Management District (SFWMD) project site soil was P-deficient Immokalee fine sand (Arenic Alaquods), a typical Spodosol with distinct A, E, and B_h horizons. This was the same soil series used in Phase I studies (A horizon only), and was expected to be responsive to P additions. Selected chemical properties of the Immokalee soils are presented in Appendix Table A-1. The materials used in the study were (1) a "low-soluble P" (~1.4 g kg^{-1}) biosolids from Pompano Beach, Fla.; (2) a "high-soluble P" (5.2 g kg^{-1}) biosolids from Boca Raton, Fla.; (3) a poultry waste material from an egg-laying operation (Tampa Farms) in Indiantown, Fla.; (4) triple superphosphate (TSP) fertilizer; and (5) an Al-WTR from Bradenton, Fla. Poultry manure/waste and Al-WTR were also used in our Phase I studies. Chemical properties of the organic sources of P and the WTR are given in Appendix Table A-2. The project experimental design was a 48-plot randomized block, consisting of three blocks, each evaluating four P-sources (two biosolids, TSP, and poultry manure), surface-applied at two rates (P-based and N-based), with and without Al-WTR. An extra plot per block was left untreated to serve as the control. Established bahiagrass (*Paspalum notatum* Flugge) was the test crop. Thus, there were a total of 51 plots, each 0.35 acre (0.14 ha) effective treatment area.

Differences in plant available N supplied by the various P-sources applied at the targeted P-based rate (35 lbs P A^{-1}, 39 kg P ha^{-1}) were equalized with variable additions of NH_4NO_3 to attain a uniform (recommended) N rate of (160 lb N A^{-1}, 179 kg N ha^{-1}). The targeted N-based rate of amendments provided the same uniform N rate of 160 lbs N A^{-1}, but different P loads (~80–160 lb P A^{-1}, ~90–180 kg P ha^{-1}) owing to the different N and P concentrations of the amendments. Amendments were applied once, in May 2003, and not reapplied before the second season of 2004.

The study pastures were mowed to a uniform height one week before the surface application of materials. Soil samples from the A, E, and B_h horizons were obtained June 25, 2003, and again on approximately 6-month intervals until project termination in December 2004. Bahiagrass was harvested three or four times each year and analyzed for P as well as yield estimates (Plank, 1992). Each plot was equipped with runoff collectors and wells with the intent of quantifying runoff and leached P (details provided at http://www.residuals.ws/).

The SFWMD experiment was partially designed to yield phytoavailability data appropriate to validating the greenhouse data from Phase I. However, the main objective was to document problems (P losses to surface and ground water) associated with residuals applications as a function of P-sources and Al-WTR amendment (to minimize the problems). Multiple biosolids sources and rates, which would have best validated the greenhouse data, were minimized (only two sources and two rates) to accommodate WTR rates and a poultry waste of local interest. Further, weather (hurricanes) and logistic problems in the spring of 2004 delayed and, ultimately, prevented N-fertilizer additions to the plots in 2004. This lack of supplemental N (to remove N-limitations to P uptake and to account for differences among the P-sources in N availability) seriously confound interpretation of the 2004 season yield and P uptake data.

2.1.2 Additional Field Studies

The project team was aware of several published and unpublished field studies involving biosolids as nutrient sources for a variety of crops. The team anticipated being able to utilize data from the studies in the validation of greenhouse determinations of relative phytoavailability (Phase I work). Almost all of the field studies known were originally designed to focus on biosolids-N issues, but extraction of P data was expected to be straightforward. Field studies by colleagues in the United Kingdom focused on P issues in manure- and biosolids-amended soils, and were expected to yield especially valuable information.

Extraction of P data pertinent to validation of greenhouse results proved more difficult than expected. Because many experiments were so heavily focused on N issues, treatments commonly were designed to logically vary residuals-N loads. Associated residuals-P loads varied, but differed when different residuals with dissimilar N and P contents were evaluated at the same N load. Similarly, other growth affecting factors (e.g., other nutrients, salts, metals) were not controlled, and confound data interpretation. Fertilizer control treatments (when more than one was included) sought to establish N loads similar to residuals-N loads and ignored P. Thus, no P phytoavailability comparisons to fertilizer-P were possible. In many regions of the United States and the developed world, agricultural soils test "high" or "excessively high" in available P as a result of historical fertilization and waste management practices. Crops growing on these soils are unlikely to respond to further P additions so differential plant responses to differences in residuals-P forms and solubility are muted, and relative phytoavailability determinations are difficult to identify. Most of the field studies identified were conducted on soils with large P-sorbing capacities—a common characteristic of all but sandy coastal plain soils in the United States. Coastal plain soils are inherently low in Fe- and Al-based minerals that sorb P, allow significant P leaching, and are often P-deficient. Differences in residuals-P forms and solubility are best determined in such soils, whereas differences are usually masked in soils with large P-sorbing capacity.

In light of the issues discussed above, several researchers were contacted to clarify details of published reports or to inquire about data and experimental conditions used in unpublished studies. The conversations led to the conclusion that none of the studies could provide unambiguous data appropriate for validating the greenhouse phytoavailability results. The United Kingdom study (Smith et al., 2003), however, was done by researchers familiar with the confounding effects of high background soil-P levels, the need for fertilizer-P controls, supplying sufficient N to account for variations in residuals sources, etc. Ten biosolids, including mostly conventionally treated materials, but also two BPRs—one Fe-dosed and one not dosed—were included. Materials were surface-applied to field plots at five rates, and the test crop (ryegrass) was harvested three times in the one-year study. The method used to quantify the relative phytoavailability of biosolids-P was unconventional, and has not been subjected to peer review, but resulted in data appropriate for our purposes. The study was discovered during a visit by the project team to England to improve mutual awareness of biosolids-P research.

2.1.3 Additional Greenhouse Data

Lacking sufficient field data to fully validate Phase I results, additional greenhouse data were collected. One completed study (Ebeling et al., 2003) compared the phytoavailability of P in various animal products with the Wisconsin BPR material used in Phase I work and fertilizer-P. The study also compared phytoavailability estimates of materials thoroughly mixed with pure sand, with sand plus small amounts of a local soil, and with soil alone. The study was well designed and went to great lengths to remove confounding factors associated with variable nutrient levels supplied by the P-sources other than P. Details of the experimental design are given in Ebeling et al. (2003).

Another completed study (Sarkar and O'Connor, 2005) compared biosolids-P responses in two Florida soils with high native P contents. Three pelletized biosolids, including two used in Phase I work (Baltimore and Largo materials), and fertilizer-P were thoroughly mixed with the soils at four rates and bahiagrass was grown for 4 months in a pot study without leaching. Details are given in Sarkar and O'Connor (2005).

Smith and colleagues shared information from unpublished greenhouse and published laboratory incubation studies that addressed various aspects of biosolids- and manure-P phytoavailability (Smith, 2003). Most of the data for conventional biosolids were from older studies, but recent laboratory studies focused on the unique P-extractability of BPR and heat-dried materials. Details of the studies are given in Smith et al. (2002, 2003).

2.1.4 Soil and Plant Analysis

Soil samples from both the field and University of Florida greenhouse studies were analyzed for total P, Fe, Al, Ca, and Mg by inductively coupled plasma atomic emission spectroscopy (ICP) following U.S. EPA method 3050B digestion (U.S. EPA, 2000). Total C and N concentrations were determined in the amendments by combustion using a Carlo Erba (Milan, Italy) NA-1500 CNS analyzer. The oxalate-extractable Fe, Al, and P were also determined by ICP after extraction at a 1:60 solid:solution ratio (McKeague et al., 1971). Water-extractable P was determined by extracting the soil samples with deionized water at a ratio of 1:10 soil:solution ratio for 1 h (Sharpley and Moyer, 2000). The P concentration in the extract was analyzed by colorimetrically using the method of Murphy and Riley (1962). Mehlich 1-P was determined on the samples by 5 min shaking of the soil samples with 0.0125 M H_2SO_4 in 0.03 M HCl solution at a ratio of 1:4 soil:solution ratio (Hanlon et al., 1997). Extracts were filtered through Whatman 42 filter paper and analyzed colorimetrically with the Murphy and Riley method (1962). The soil iron (Fe)-strip P (ISP) involves reacting the samples with Fe-impregnated (0.65 M $FeCl_3$ in 0.6 M HCl) filter paper (Chardon et al., 1996). The P adsorbed is then extracted with 0.1 M H_2SO_4, and the extractable P analyzed colorimetrically with the

Murphy and Riley method (1962). Standard quality assurance/quality control protocols were observed during the sample collection, handling, and chemical analysis.

Grasses from the field and greenhouse studies were harvested approximately monthly, dried, weighed, digested, and analyzed for P (Plank, 1992). Phosphorus uptake was calculated as the product of yield times P concentration and was related to P-source, P-rate, WTR treatment, and extractable soil P concentrations.

The soil and plant data were statistically analyzed as a factorial experiment with a randomized complete block design using the general linear model of the SAS software (SAS Institute, 1989). The differences among the treatment means were separated using a single degree of freedom orthogonal contrast test. All statistical analysis tests were done at a significance level of 5%.

2.2 Rainfall Simulation Studies

2.2.1 P Source Collection and Characterization

Manures, biosolids, and commercial fertilizers collected for the rainfall simulations studies were analyzed for percent total solids, volatile solids, water-extractable P, and total P, Fe, Al, and calcium (Ca) concentrations. Total elemental content (total P, Fe, Al, Ca) of the P sources was determined using acid digestion (EPA method 3051) followed by elemental analysis via ICP. Total N was determined by combustion at 1100 °C using an Elementar CNS Analyzer. Solids content (103–105 °C, 15 h) and pH (1:2 P source to distilled water) were determined by standard methods. All analyses noted above were performed in duplicate and rerun when results differed by more than 15%.

For water-extractable P analysis, distilled/deionized water was added to a 0.5 g (dry weight equivalent) of P-source (as received) to give a 1:200 solid:solution ratio. Samples were agitated for 1 h at room temperature (end-over-end shaker at ~15 rpm), centrifuged (15 min at ~1250 rpm), and vacuum-filtered through pre-wetted 0.45-µm filter paper. Samples were prepared for ICP analysis by adding 0.5 mL HCl (1:1) to 24.5 mL filtrate. Two replicates of water-extractable P extractions were performed.

2.2.2 Soil Collection and Characterization

For two of the indoor rainfall simulations, the surface horizon (0–-20 cm) of Hagerstown silt loam (fine, mixed, semi-active, mesic Typic Hapludalfs) soil was collected, field sieved (6.4 mm), air dried, and mixed for use in the runoff experiments. Total elemental concentrations were determined using acid digestion (U.S. EPA method 3051) followed by elemental analysis via ICP (U.S. EPA method 6010). Mehlich-3 extracts (Mehlich, 1984) were also analyzed using ICP. For the WTR amendment rainfall simulation, Immokalee fine sand (previously described) was used.

2.2.3 Indoor Rainfall Simulations

The rainfall simulations generally followed the National Phosphorus Research Project indoor runoff box protocol (National Phosphorus Research Project, 2001). Stainless steel runoff trays (20- × 100- × 7.5-cm deep) were packed with 5 cm of the Hagerstown silt loam and the P sources were uniformly applied to the surface of the soil. Three trays were prepared for each treatment and three trays with unamended soils served as controls. The trays were first wetted to field capacity and allowed to air dry for 24 h. The P sources were then applied and the rainfall events were conducted 2- and 4-d after P-source application. Trays were sloped at 3% and rainfall was delivered at 71 mm hr^{-1}. For each event, the initial 30 min of runoff was collected and a subsample was immediately filtered (0.45 μm) and acidified (pH < 2). Total dissolved P was determined on the filtered samples by ICP. Total P was measured on the unfiltered runoff samples by ICP following digestion using EPA method 3051 (U.S. EPA, 2000). The difference between total P and total dissolved P for any sample was assumed to represent particulate P. Two or three rainfall-runoff simulations were performed for each experiment, but discussion focuses on the first rainfall event since most P is lost in the initial rainfall event for biosolids-amended soils (Sims et al., 2003). Because differential effects of P sources are generally the same based on runoff P loads (mg) or concentrations (mg L^{-1}) (Kleinman et al., 2002a), the researchers have, like other recent rainfall simulation studies (Kleinman and Sharpley, 2003), confined the discussion to runoff P concentrations.

Rainfall simulations with the Immokalee fine sand required a modification of the typical runoff box protocol design to account for P leaching. A second box was positioned under the first in a double-decker design to collect and quantify P leached from the top box. The design allowed collection of runoff and leachate P simultaneously. Soils were pre-wetted to near saturation before P sources were surface applied. The amended soils were allowed to equilibrate for one day before the first of three rainfall events on an alternate day schedule. Thirty minutes of runoff generated during each rainfall event were collected from each box, and the volumes recorded. Leachate was collected continuously during each rainfall event, and began before runoff began. Runoff and leachate was subjected to the same analyses described for the Hagerstown soil, but also included a determination of ISP (Van der Zee, 1987) as a measure of bioavailable P (Sharpley, 1993a,b).

2.2.4 Field Rainfall Simulation

The research site was located at the Pennsylvania State University's Larson Research Farm in Rock Springs, Pa. The site had an 8% slope and an established stand of mixed grasses (orchard grass, timothy, fescue). The site was located on a Buchanan soil (fine-loamy, mixed, semi active, mesic Aquic Fragiudults). Buchanan soils have slow permeability, a fragipan at 50+ cm, and are moderately well drained. The site was selected after preliminary soil analysis determined that the land was suitable for biosolids application under Pennsylvania regulations. The grass was cut to 10 cm and the clippings were removed from the plot area.

Runoff plot dimensions were 2 m long × 1 m wide, directed down slope. Each plot was adjoined to another plot for analysis replication. Metal borders (0.2 cm thick and 15 cm high) were installed 5 cm above the ground to isolate the surface runoff inside the plot. A runoff gutter (equipped with a canopy to exclude direct rainfall) was installed at the lower edge of each plot, directing the runoff to a collection point. This study employed six "paired" plots (see plots A + B, Figure 2-1).

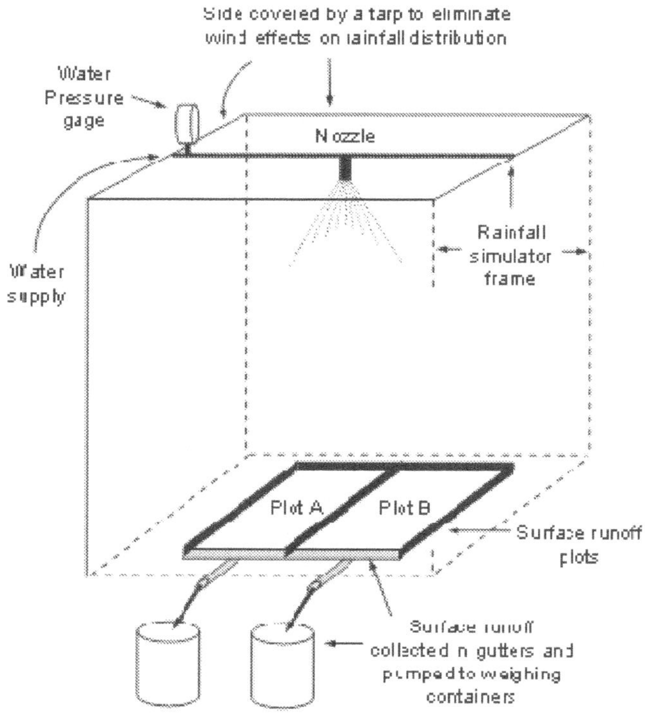

Figure 2-1. Portable Rainfall Simulator Design.

Ten soil cores were collected (and bulked) for a sampling depth of 2.5 cm from outside each plot before the first rain event to provide a soil test P (STP) of the field. After the final rain event, six soil cores (2.5 cm) were collected from inside each plot and bulked to obtain STP of each plot after P source application. All soil samples were air dried and sieved (2 mm). The soil was analyzed for Mehlich-3 P by shaking 1 g of soil with 10 mL of Mehlich-3 solution (0.2 *M* CH_3COOH + 0.25 *M* NH_4NO_3 + 0.015*M* NH_4F + 0.013 *M* HNO_3 + 0.001 *M* EDTA) for 5 min. The supernatant was filtered (0.45 μm) and P in the filtrate was analyzed by ICP.

To observe physical characteristics and P content of the P source materials over time, randomly selected samples from each of the six P sources were collected from inside the plots three times during the period between the third and fourth rain simulations. A piece of paper (7.6 cm^2) was dropped from a height of approximately 3 feet towards the rear of each paired plot and the P source sample was collected from the area underneath the paper. Care was taken to exclude soil from the sample. Each sample was analyzed for percent solids (determined gravimetrically), water-extractable P, and total P.

Kleinman et al. (2001) described the portable rainfall simulator used by the USDA Agricultural Research Service and employed in this study. The simulator had been closely matched with natural rainfall and offers a control of variables such as rainfall intensity and uniformity. Figure 2-1 shows the outdoor rain simulator design and positioning over the field plots. An aluminum frame held the nozzle, piping, wiring, and pressure gauge. A TeeJet™ ½ HH SS 50 WSQ nozzle (Spraying Systems Co., Wheaton, Ill.) was positioned approximately 3.1 m above the center of the plots. Water flow rate (233 mL s^{-1}) was established by adjusting a pressure regulator, adjacent to the nozzle. The flow rate was measured before each simulation to ensure consistent rainfall intensity (71 mm h^{-1}). To minimize wind disturbance, tarps were placed around the frame.

Source water was collected from a spring-fed irrigation pond at the foot of a forested hill slope with no upslope agricultural activity. The source water contained negligible (<0.06 mg L^{-1}) concentrations of both dissolved inorganic and total P.

Four rain simulations were performed at each plot: three simulations were conducted at 24-h intervals (Days 0, 1, and 2) and a fourth simulation was completed 22 days after the initial simulation. The first rainfall event (Day 0) was intended to set field capacity soil moisture conditions for the following rainfalls, as well as to provide a background soil P runoff concentration for the field. Manures and biosolids were manually surface-applied at a rate of 100 kg total P ha^{-1} to each plot after the initial (Day 0) rainfall event. All treatments were conducted in duplicate. Simulated rainfall was applied until 30 min of runoff was collected from each treatment. Runoff was collected in ~15-L containers every 5 min and weighed to determine mean runoff flow rate and total runoff volume. A bulk sample of total runoff was collected from each plot for analysis.

Natural precipitation occurred during the 3 weeks between the third and final simulation. Runoff from natural rainfall was collected from only one randomly chosen plot (site 3, plot B), because of lack of resources. Collection was similar to that used for the rainfall simulations. Total runoff was weighed at the end of each precipitation event.

2.2.5 Runoff Collection and Analysis

Collected runoff was analyzed for total runoff mass, total solids, volatile solids, total P, and total dissolved P. The various analyses were performed on aliquots taken from a single composite sample representing the 30 min of runoff from each plot. Samples were refrigerated (1–5 °C) in sealed plastic bottles until analyses were conducted.

Total P was determined following a modified semimicro-Kjeldahl procedure on unfiltered runoff water (Bremner, 1996) and measured colorimetrically at 712 nm by an ascorbic acid method (Murphy and Riley, 1962). Subsamples of runoff from each plot were vacuum-filtered through pre-wetted 0.45-μm filter paper within an hour after runoff collection and acidified to pH <2. The filtered samples were analyzed for total dissolved P by ICP analysis (U.S. EPA method 6010B). Runoff water extracts analyzed colorimetrically for dissolved reactive P (Murphy and Riley, 1962) generally results in lower P values than runoff water analyzed for total dissolved P by ICP (Maguire and Sims, 2002a; Sims et al., 2003).

Particulate P (PP) was calculated as the difference between total P (TP) and soluble phase P (TDP) (PP = TP – TDP). Particulates associated with erosion: total (RTS), fixed (RFS), and volatile (RVS) solids in runoff were analyzed according to standard protocol (Clesceri et al., 1998). Runoff total solids were determined by drying ~150 mL runoff samples at temperatures of 100–105 °C. Runoff fixed solids were determined by heating samples to an additional 550 °C $\pm$ 50 °C. Runoff volatile solids are those lost on ignition at the high temperatures, calculated by RVS = RTS – RFS.

Natural rainfall was analyzed for solids (Clesceri et al., 1998), total P and total dissolved P by colorimetric analysis. Note that, for the natural runoff event, only total solids and dissolved inorganic P were measured. Total P was determined by Kjeldahl procedure on unfiltered runoff water (Bremner, 1996) and measured colorimetrically (Murphy and Riley, 1962). Dissolved inorganic P was determined on a filtered sample by the molybdate blue method with ascorbic acid as a reductant (Murphy and Riley, 1962).

CHAPTER 3.0

RESULTS OF FIELD STUDIES AND RAINFALL SIMULATIONS

3.1 Introduction

Two overriding objectives guided the research and form the organization framework for presentation of results. The first objective was validation of the P phytoavailability information gathered through laboratory and greenhouse experiments in Phase I (O'Connor et al., 2002). In cooperation with the SFWMD, a major project was undertaken to evaluate plant, soil, and water quality implications of land applying biosolids-, manure-, and fertilizer-P in a pasture setting. Plant P uptake was quantified as well as various soil measures of P extractability, and the effect of WTR co-application was addressed. Additional field and greenhouse studies are also discussed that address the issue of P phytoavailability.

Building on experience gained from initial rainfall simulation experiments (Brandt, 2003), a series of rainfall simulation studies was conducted to quantify and compare P runoff losses from various surface-applied P sources. This served as the second major project objective. Five rainfall experiments were conducted. A **multiple P source simulation** (Section 3.3.1) was conducted using 24 biosolids, five manures, two fertilizers, and one food processing residual. A **variable P loading rainfall simulation** (Section 3.3.2) was conducted where the P sources were applied to give a common plant-available N rate to the soil. A **WTR amendment simulation** (Section 3.3.3) was conducted to quantify the impact of co-application of WTRs and P sources. A **field rainfall simulation** (Section 3.3.4) was conducted to evaluate P losses from grassed plots under field conditions. Finally, an **application method simulation** (Section 3.3.5) was conducted to evaluate management methods in P runoff loss.

3.2 Validation of P Phytoavailability

3.2.1 SFWMD Field Study

3.2.1.1 P Uptake

Data from the two-year SFWMD field study on established bahiagrass pastures are summarized in Figures 3-1 (2003 growing season) and 3-2 (2004 growing season). Data details are provided in Appendix Table A-3. Data for 2004 reflect residual treatment effects, as amendments were applied only in May, 2003.

Analysis of variance (ANOVA, Appendix Table A-4) of the plant data revealed no effect of P-source or P-source rate on total grass yields or total P uptake in either growing season. Application of WTR also had no effect on total grass yields in either year, but significantly reduced P uptake and weighted average P concentrations of grass tissue in both years. Despite being decreased by WTR treatment, P concentrations of tissues remained well above the 1 g P kg^{-1} value regarded as minimally sufficient for bahiagrass. Thus, no plant P deficiencies occurred as a result of WTR application, and forage value of the grass to grazing cattle was not severely affected. Application of 1% (10 T A^{-1}, 22.4 Mg ha^{-1}) WTR essentially masked P-source and P-source rate effects on plants.

Of particular interest to waste managers was the similar plant yield and P uptake values in both years by grass exposed to any P-source applied at N-based rates in the presence of WTR, and P-sources applied at P-based rates in the absence of WTR. The data suggest that WTR applications can counter elevated soil P concerns associated with N-based rates of residuals (manures or biosolids—even those with high soluble P concentrations) without negatively affecting the agronomic value of the grass.

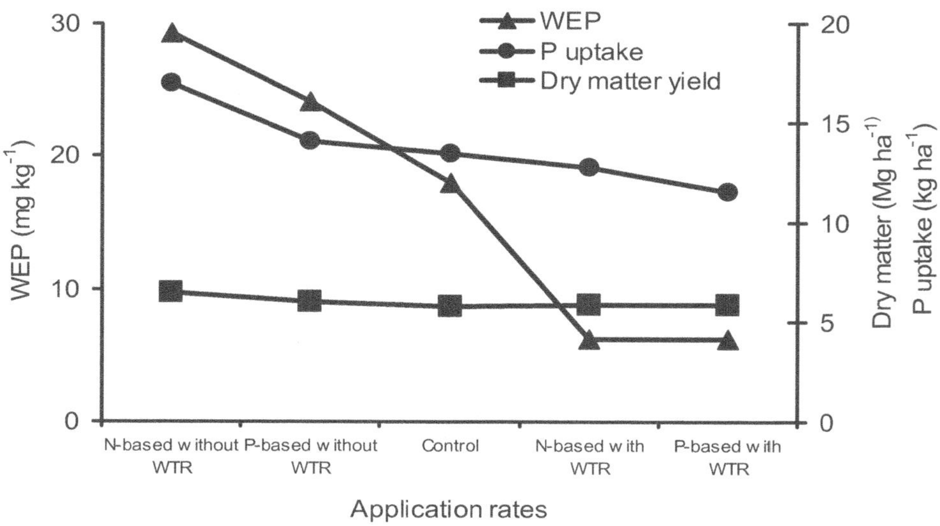

Figure 3-1. Variation of Water-Extractable P (WEP), Dry Matter Production, and P Uptake Values with Treatment Applied in the SWFMD Study in 2003.

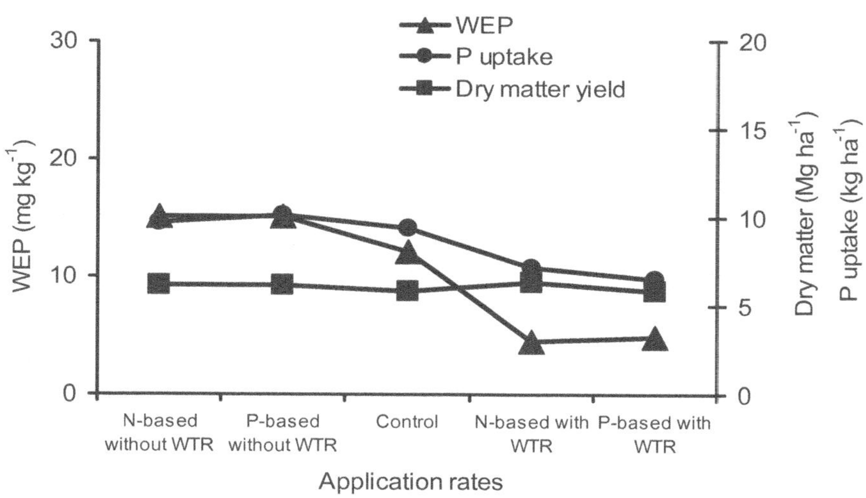

Figure 3-2. Variation of Water-Extractable P (WEP), Dry Matter Production, and P Uptake Values with Treatment Applied in the SWFMD Study in 2004.

Water-extractable P of the surface (0–5 cm depth) soil at the beginning of the experiment was largely unaffected by P-source or P-source rate, but strongly affected (decreased) by WTR application (Figure 3-1). Treatment effects on water-extractable P were expected to correlate well with treatment effects on P uptake and water quality based on work from Phase I (O'Connor et al., 2004; Elliott et al., 2002). The lack of WTR effect on agronomic grass value, despite clear effects on water-extractable P values, can be explained in three ways:

1. WTR (and other) treatment effects were limited to shallow soil depths (0–5 cm), whereas established bahiagrass has roots extending into the B_h (spodic) horizon, typically, at ~40 cm. Most roots occurred in the top 15 cm of soil, but treatment effects were largely restricted to the 0–5 cm depth because treatments were surface-applied and not incorporated into the soil. Roots in the unaffected rooting volume of soil likely supplied sufficient P to the grass to maintain agronomic value.

2. High variability in soil P concentrations across, and with depth within, the plots may have masked P-source and P-source rate effects. Bahiagrass pastures in the region have repeatedly failed to respond to P-fertilizer trials despite very low STP values. Soil samples utilized for soil testing (typically, 0–15 cm deep) may not adequately represent the extreme variability in actual phytoavailable P in the area soils.

3. Plants may be able to utilize some of the P sorbed by WTR, in addition to P remaining in water-extractable form after WTR treatment. Plants utilize a variety of mechanisms (e.g., release of organic chelating agents, lowering of rhizosphere pH) to access P held to soils that is not water extractable (Kleinman et al., 2000).

Phosphorus uptake data from treatments not involving WTR additions are appropriate for determining the relative phytoavailability of different P-sources and for validating greenhouse data generated in Phase I work. Data from the first growing season were regarded as most useful, as N availability in the second growing season was likely inadequate and uncontrolled. Additionally, the plots were frequently inundated during the 2004 season because of unusually high hurricane activity. The first growing season data were further pared to utilize only those plots receiving nearly equal P loads and uniform and adequate N loads – the P-based rates. Given the limited number of data points, relative P phytoavailability (RPP) was estimated as point estimates, rather than the slopes (Phase I work), of P uptake response curves to P loads. Thus, the ratio of P uptake from plots treated with an organic source of P to the average total soil P concentration over the growing season for those plots was divided by a similar ratio for fertilizer (TSP)-treated plots, times 100:

$$RPP = \frac{\text{P uptake}_{\text{Organic source}} / \text{Average soil P conc.}_{\text{Organic source}}}{\text{P uptake}_{\text{TSP}} / \text{Average soil P conc.}_{\text{TSP}}} \times 100 \qquad (3.2.1.1\text{-}1)$$

The resulting RPP values from the field study (shown in boldfaced type) and greenhouse data from Phase I work are presented in Table 3-1, along with water-extractable P values for each P-source.

Table 3-1. Relative Phytoavailability (RPP), Category of Relative Phytoavailability, and Water-Extractable P (WEP) of P-Sources Used in Greenhouse and Field (Boldfaced) Studies.

Source	RPP (%)	Category of RPP	WEP (g kg^{-1})
TSP	100	High	21
Largo cake	74	High	8.8
Wisconsin cake	104	High	6.2
Boca Raton biosolids	**85**	**High**	**5.2**
Poultry waste	**55**	**Medium**	**3.5**
Tarpon Springs cake	31	Medium	3.4
Largo pellets	48	Medium	2.6
Untreated aerobic cake	54	Medium	1.7
Pompano biosolids	**59**	**Medium**	**1.4**
Baltimore cake	34	Medium	0.86
Philadelphia cake	37	Medium	0.57
Untreated anaerobic cake	26	Medium	0.29
Baltimore pellets	9	Low	0.17
Philadelphia compost	7	Low	0. 05

The estimates of relative phytoavailability generally track well with the relative water extractable P values of the P-sources. In particular, the water-extractable P value for the Boca Raton material is the greatest of the organic sources of P used in the field and had the greatest RPP value. Indeed, the RPP is similar to values for the "high" phytoavailability category of biosolids examined in Phase I work. The Boca Raton material is not produced by a BPR process as were the other two "high" phytoavailability materials identified in Phase I work. Nevertheless, the treatment plant operator (Welling, 2005) offered that the "high rate activated sludge" process apparently allows similar "luxury uptake" of P (representative of BPR processes). The resulting material has a high water-extractable P value and high phytoavailability. The Pompano material has a lower soluble P concentration, similar to biosolids categorized as "moderate" phytoavailability in Phase I work. Thus, Pompano biosolids-P mimics most conventionally processed biosolids, with RPP values similar to the U.S. EPA, Canadian, and UK regulators' average (default) value of 40–60% of fertilizer-P.

Phytoavailability of manure-P is often regarded as being about equal to fertilizer-P (Martz, 2002), but is logically expected to vary with animal type and to the extent that bedding, or other solids, are included in the final product. As with biosolids, most manure-P is inorganic, but the percentage of total manure-P that is water-extractable is usually greater than percentage of water-extractable P values of biosolids-P (Brandt et al., 2004) because metal concentrations are lower in manures.

Phase I work included three animal manure products (one dairy manure and two poultry wastes) in the greenhouse studies. Relative phytoavailability determinations were confounded by high soil pH values associated with high manure application rates and subsequent poor growth of the acid-tolerant bahiagrass. Similar problems did not arise in the field study owing to the lower P-based application rates. Comparisons of the RPP value for the poultry waste in the field to Phase I greenhouse work, therefore are problematic. The water-extractable P value for the poultry waste used in the field study was nearly the same as that for the Pompano biosolids, and within the "moderate" range of materials examined in Phase I work. The poultry waste RPP value is similar to the Pompano biosolids and suggests moderate phytoavailability, less than traditional estimates for manures (Martz, 2002). Significant amounts of bedding material were present - in addition to the layer-manure - which may have reduced poultry waste P phytoavailability.

3.2.1.2 Soil Impacts

Results of Phase I work showed that total P concentrations of manures and biosolids are unreliable measures of labile (phytoavailable or leachable) P. Similarly, total P in manure- or biosolids-amended soils and a common soil test (Mehlich-1) did not predict P uptake well, especially in the presence of added WTR. Greenhouse results of Phase I, however, suggested that water-extractable P and ISP could be effective predictors of P uptake in soils amended with manures, biosolids, and/or WTR. Further evaluation of the efficacy of the soil tests under field conditions was recommended.

Soil samples [0–5 cm depth (both years) and 0–15 cm (2004 season only)] taken from the SFWMD project site were subjected to various measures of P extractability, including total P, Mehlich-1, ISP, and water-extractable P. The resulting extractable P values of the shallow soil samples (0–5 cm) were then correlated with bahiagrass yields in each season (Appendix Table A-5). Data for both years were also combined to yield correlation coefficients across years. The data were separated into two groups: data representing all treatments and data for treatments not involving WTR application, to highlight the differential ability of the extractants to predict P uptake in the presence of WTR.

The analysis was expected to be of limited value because only shallow (0–5 cm) surface soil characteristics were considered, whereas plant uptake is typically characterized with samples from the 0–15 cm depth that represent more of the rooting zone. Indeed, most correlation coefficients were low (<0.5), although a few were greater, and several were statistically significant (Appendix Table A-5). In general, water-extractable P correlated best with P uptake in each season, across seasons, and in both presence and absence of WTR. ISP correlated nearly as well with P uptake as water-extractable P, but not as well for treatments not involving WTR. Total P correlated reasonably well with P uptake for those treatments not involving WTR, but was not a reliable measure of plant available P when the WTR-related treatments were included. Thus, results from the field study are consistent with greenhouse results (Phase I work), but the soil sampling scheme dictated in the field study did not allow rigorous testing of the Phase I results.

Measures of surface (0–5 cm) soil water-extractable P are expected to better indicate P environmental lability—runoff (Pote et al., 1999; Vadas et al., 2005) and leaching (Kleinman et al., 2003) than phytoavailability. Nair et al. (2004) showed that soil water-extractable P values tracked well with the degree of P saturation (DPS) values for Florida soils. The DPS value was developed in The Netherlands to quantify the risk of P leaching in sandy soils (Breeuwsma et al., 1995). The Dutch investigators showed that when the DPS value of a soil exceeded ~25%, the risk for loss of subsequently added P increased sharply. Nair et al. (2004) showed that the critical DPS value (identified as a "change point" in the relationship between water-extractable P and DPS) was ~20% for Florida soils.

Values of water-extractable P for surface (0–5 cm) and "plow-layer" (0–15 cm, 2004 season only) samples of the various treatments utilized in the SFWMD field study over time are shown in Figure 3-3; detailed information is provided in Appendix Table A-6. The DPS values

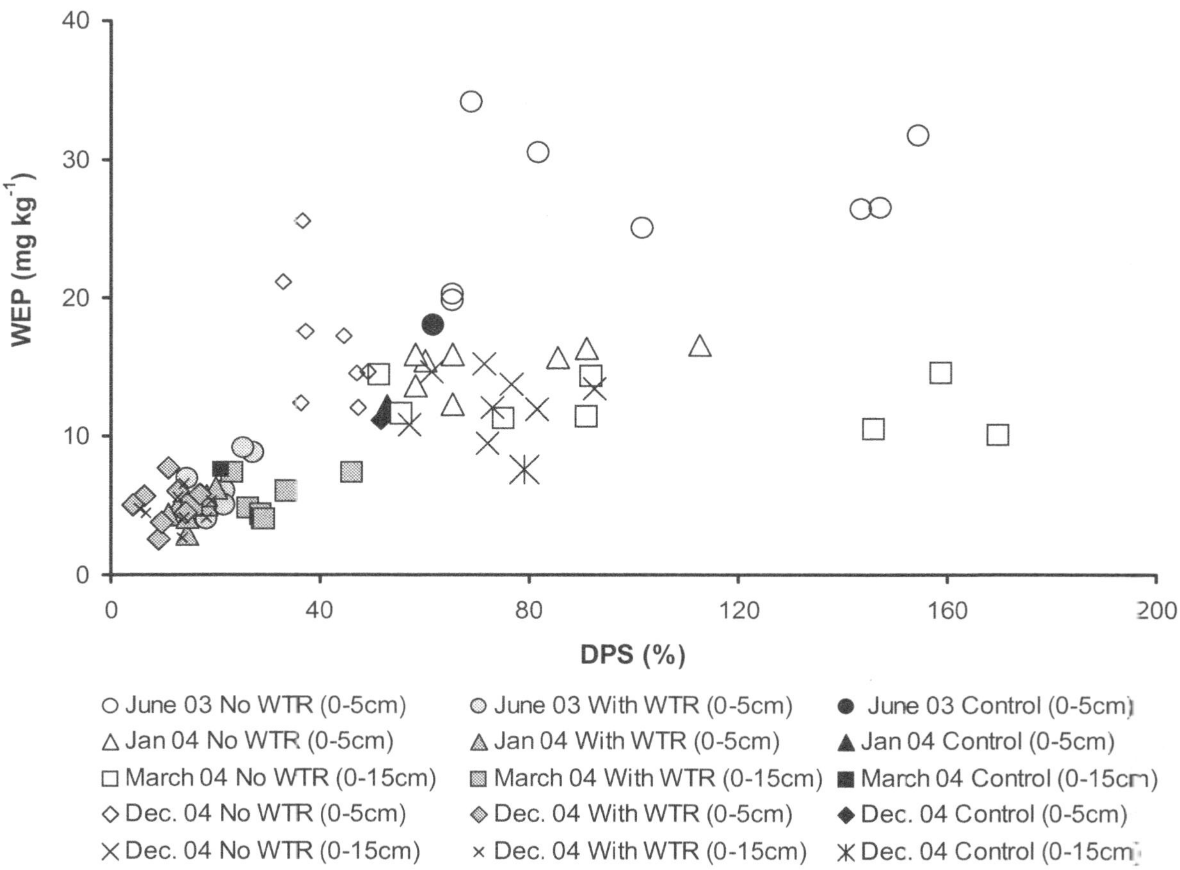

Figure 3-3. Water-Extractable P as a Function of Degree of P Saturation (DPS) Over Time for Two Soil Depths in the SFWMD Study.

of all soils representing P-source and P-source rate treatments, except those representing WTR addition, exceeded the critical value of 20% throughout the two-year experiment. The data imply that any P-source, applied at any rate, results in high risk of P loss to runoff (0–5 cm data) or leaching (0–15 cm data). The various P-sources affected DPS values differentially, reflecting water-extractable P values of the P-sources (Appendix Table A-2), but all DPS values for all sources exceeded 20%. Treatment with WTR reduced all DPS values to ~25%, or less, and masked P-source and P-source rate treatments, implying dramatically reduced risk of P loss.

The combined plant and soil data from the field study support the contention (Phase I work) that WTR can control soluble P problems associated with high (N-based) rates of even high soluble P materials without adversely affecting bahiagrass yields or nutritional value to grazing cattle. In particular, the field data support the use of Al-WTR applied at 10 T A^{-1} (22.4 Mg ha^{-1}) to reduce P loss (runoff and leaching) from residuals-amended soils to losses similar to those expected when residuals are applied at P-based rates.

3.2.2 Additional Field Studies

Smith et al. (2003) utilized samples from a field study focused originally on N availability from 10 biosolids (Morris et al., 2003). Materials included examples of liquid and dewatered digested, thermally dried, composted, and lime-stabilized biosolids as well as liquid thermally hydrolyzed and prepasturized products. Two dewatered digested products from nutrient-removal processes—one by Fe-dosing and one by BPR—were also included. Materials were applied at five rates to establish a range of N loads. A series of fertilizer-N (NH_4NO_3) control plots established a N load gradient (0–120 kg N ha^{-1}) similar to the biosolids-N loads. One fertilizer series withheld added P and another included P at 21.8 kg P ha^{-1} (as TSP) across all N plots.

The complex experimental design allowed a statistical manipulation of the data to account for "background" P uptake on the control treatment (without P addition) in response to increased yields associated with N application. This background removal was then subtracted from the P removal estimated for the biosolids and fertilizer-P to yield estimates of P removal directly attributable to the P-sources. The values could then be expressed as P removal per mass of P added, and used to calculate the relative phytoavailability of biosolids-P to fertilizer-P.

The fields used in the study apparently contained abundant available P, as the linearized relationships between P uptake and yield for the fertilizer N and fertilizer N plus fertilizer P were barely distinguishable (Smith et al., 2003, Figure 2). The minimal plant response is common in soils containing abundant available P, and such conditions are common in the United States and Europe where historical fertilizer and waste applications have increased soil-P concentrations. Such soils are not ideal for the differentiation of the relative phytoavailability of various P-sources, as differences are muted (de Haan, 1980). The soils, however, are important and practical test sites.

Despite the modest responsiveness of the soil and the data manipulations used, Smith et al. (2003) provided valuable estimates of biosolids-P phytoavailability. Smith et al. reached four conclusions:

1. Conventional [e.g., mesophilic anaerobic digestion (MAD), with or without Fe-dosing] biosolids were 53–59% as bioavailable as fertilizer-P;
2. Liquid products (MAD, thermally hydrolyzed MAD, and thermophilic aerobic MAD biosolids) were 200–300% as bioavailable as fertilizer-P;
3. A BPR biosolids was 125% as bioavailable as fertilizer-P; and
4. Two thermally dried biosolids and a lime-stabilized biosolids were 95–106% as bioavailable as fertilizer-P.

Conclusions 1 and 3 are consistent with previous United Kingdom and United States work that conventional produced/treated biosolids have RPPs about 50–60% of fertilizer-P and that BPR biosolids-P is very similar to fertilizer-P. Conclusions 2 and 4, however, are difficult to explain, and even conflict with laboratory estimates of phytoavailability in thermally treated products by Smith et al. (2002). Apparently, liquid biosolids-P contains appreciable amounts of immediately available (soluble) P that is even more readily available than fertilizer-P (applied at equal P rates) in a one-year study. The researchers have no explanation for the ready availability of thermally treated materials noted by Smith et al. (2003), and the results conflict with reduced phytoavailability noted in Phase I work. The researchers concur with Smith et al. (2003) that additional, long-term field studies are needed to clarify the issue, and to devise appropriate advice for those involved in the land application of biosolids. Unfortunately, the UK field study was discontinued because of a lack of funding.

3.2.3 Additional Greenhouse Studies

Additional greenhouse studies of RPP have been completed and published since the completion of Phase I work. Ebeling et al. (2003) compared the phytoavailability of P in various animal manures with that of the Wisconsin BPR material used in Phase I work and fertilizer-P, using wheat as the test plant. In pure sand culture, the BPR material resulted in equal wheat yields as fertilizer-P, but P uptake was about 170% that measured in the fertilizer-P treatment. Phosphorus rates were high (101 and 202 kg P ha^{-1}) and P-sources may have altered important soil properties (e.g., pH) and the availability of other nutrients in the unbuffered sand media (Ebeling et al., 2003). A similar study using Ringwood silt loam (fine loamy, mixed, mesic, Typic Argiudolls)—likely, much better pH-buffered and with appreciable P-sorbing capacity—was also conducted. The BPR material resulted in greater wheat yield and greater P uptake than fertilizer-P, and the RPP of BPR biosolids was estimated as ~130% of fertilizer-P across P-rates. The authors emphasized that P-sources with high concentrations of extractable P may not result in correspondingly high P uptake rates because of growth medium (soil) characteristics, and that measures of P phytoavailability should be done in soils expected to be amended with the sources. The guidance was especially pertinent to the explanation of the relative phytoavailability of the manure sources. The BPR biosolids exhibited greater phytoavailability than fertilizer-P in both sand and soil cultures, and results agreed with those for the Wisconsin BPR found in Phase I work.

Sarkar and O'Connor (2005) confirmed the observations of Ebeling et al. (2003) that soil characteristics are critical to determinations of relative phytoavailability. Three pelletized biosolids (including the Baltimore and Largo materials used in Phase I work) were applied to two soils testing "moderately high" (Candler) and "very high" (Millhopper) in STP. The Millhopper soil also contained appreciably more oxalate-extractable Fe + Al (2720 mg kg^{-1}) than the Candler soil (636 mg kg^{-1}) and the soils exhibited correspondingly high and moderate P-sorption capacities. Few differences in plant responses to P-sources or P-source rates were noted in the Millhopper soil. The BPR Largo material consistently resulted in greater P uptake than fertilizer-P, but the effect was significant only at the highest (N-based) source rate. High initial STP and strong P sorption in the Millhopper soil essentially masked biosolids-P characteristics. In the Candler soil, plant responses to all P-sources and P-source rates were greater than in the Millhopper soil, but differences among P-sources were muted. At the greatest rate (N-based), the BPR Largo material again resulted in the greatest P uptake, and the high Fe + Al Baltimore material the least P uptake.

Both the Ebeling et al. (2003) and Sarkar and O'Connor (2005) studies emphasize that differences in biosolids-P relative phytoavailability are primarily expressed in soils initially deficient in P and low in P-sorbing capacity. In such situations, control of P chemistry is primarily associated with P-source differences. Even modest differences in biosolids-P relative phytoavailability are demonstrated in such soils, and P uptake is a strong function of soluble-P load. In soils testing very high in STP and/or containing appreciable P-sorbing capacity, however, control of P chemistry and plant uptake of P is primarily associated with the soil. Such soils dominate many areas of the world where biosolids are land applied. Differences in biosolids-P relative phytoavailability, and even differences in soluble P loads (biosolids application rates), are largely masked in these soils. The lone exception to this generality regarding P-source effects is BPR materials. The Wisconsin and Largo, FL BPR materials consistently resulted in greater P uptake, regardless of soil characteristics. This was particularly true when the materials were applied at rates mimicking N-based rates. Thus, most conventionally treated biosolids can be expected to exhibit agronomic values (RPP, effects on plant uptake of P) similar to fertilizer-P in many medium- to heavy-textured soils. BPR materials, however, appear to provide greater agronomic value in all soils, frequently exceeding that of fertilizer-P when applied at N-based rates.

Ongoing research being conducted by the Metropolitan Water Reclamation District of Greater Chicago (MWRDGC) promises to provide extremely valuable data to validate work conducted in the Water Environment Research Foundation project. The MWRDGC research includes both greenhouse and field studies. The greenhouse study utilizes the same P-deficient Immokalee soil used in the Florida studies, and includes multiple (six) rates of P applied as fertilizer-P (TSP) and class A and B biosolids produced by MWRDGC. The study also involves multiple croppings of the treatments to evaluate the residual value of P-sources. The field study will utilize the same two MWRDGC biosolids and TSP applied at various rates to sites in Fulton County, Illinois, where Chicago personnel have conducted extensive biosolids research for approximately 30 years.

The greenhouse study is in the final stages of the third cropping of treatments, and analyses of earlier harvests are incomplete. Data from the first cropping, however, demonstrate classic text-book variance of plant uptake of P with increasing rates of applied P. The data show clear distinctions in P uptake for the various P-sources. Uptake from the biosolids sources is obviously less than from TSP treatments, and reflects differences in the water-extractable P values of the P-sources. The study should provide additional validation of Phase I results, and will address the important issue of biosolids-P residual value.

The field study is slated to begin later in 2006 following repeated croppings to reduce STP values to levels indicative of P-responsive soils. The large-scale effort should provide rare field evaluations of biosolids-P phytoavailability of P-responsive soils.

3.3 Rainfall Simulation Studies

3.3.1 Multiple P Source Simulation

3.3.1.1 Soil and P Sources

This study investigated P losses from surface applied materials representing a wide range of amendments with variable water-extractable P values. Twenty-four biosolids materials were included, along with five manures (one swine, three poultry, one dairy), two fertilizers [diammonium phosphate (DAP) and TSP], and a mixed food processing residual. The major types of solids processing operations represented included: aerobic and anaerobic digestion, lime stabilization, heat treatment, and advanced alkaline stabilization. Two additional treatments were included to assess the influence of mixing an Al-based WTR with a BPR biosolids (Madison cake) and the swine manure on the subsequent P runoff losses. The WTR, from Manatee County

Water Treatment Plant in Bradenton, Florida, is generated in a conventional coagulation-filtration process where alum is the primary coagulant.

Properties of the materials are shown in Table 3-2. Elemental concentrations of Al, Fe, and Ca reflect chemical additions in wastewater treatment and solids processing. All biosolids had been dewatered and contained no free draining liquid. The three poultry manures were dry materials (63–79% solids content). The swine slurry (5.1% solids) and the food processing residual (3.9% solids) were applied to the soil boxes as liquids. Excluding the mineral fertilizers, total P contents of the materials were 7–38 g kg^{-1}, and the water-extractable P values were from 0.07 mg kg^{-1} (for high Fe + Al biosolids) to 9.2 mg kg^{-1} for swine manure.

To maximize the number of materials investigated, two (rather than three) replications of each treatment were exposed to a single rainfall event. Two controls (soil only) were also included, resulting in 36 treatments, each replicated two times, for a total of 72 trays. All materials were surface applied to soil (Hagerstown silt loam) trays at a total P application rate of 100 kg P ha^{-1} and allowed to preequilibrate in a moistened condition for 48 h. Simulated rainfall (71 mm h^{-1}) was then applied until 30 min of runoff was collected (see Section 2.2.3).

3.3.1.2 Runoff Results

Figure 3-4 shows the runoff total dissolved P and total P for the treatments ranked in order from lowest (control soil) to highest (poultry manure) runoff P. Although there are some noteworthy exceptions, study results generally confirm previous findings that biosolids cause lower soluble P losses than land-applied livestock manures (Brandt and Elliott, 2003). Swine and poultry manures caused greater runoff P losses than dairy manure.

Because only two replications were used to calculate the mean total P and total dissolved P levels, the statistical inferences are limited. For example, the total P values for all biosolids (except Largo pellets) and the dairy manure were not different than the controls. This included some BPR biosolids, like the Madison, Ephrata, and York biosolids. Moreover, some heat-dried products (Oceangro pellets, Baltimore Patapsco pellets, Back River pellets) had runoff total dissolved P levels not statistically different from conventional or BPR-generated biosolids. These comparisons are not consistent with results from other studies (e.g., Penn and Sims, 2002), and are believed to be artifacts of the small number of replications.

Table 3-2. Properties of the P Sources Used in the Multiple P Source Rainfall Simulation.

P Source Acronym	P Source Acronym Definition	TS (%)	pH	P Source - total concentration ($g\ kg^{-1}$) Al	Ca	Fe	N	NH_4-N	Org-N	P	WEP ($g\ kg^{-1}$)
ABC	Abington cake	21.5	7.7	11.2	42.2	17.9	59.0	8.62	50.3	26.5	0.30
ACE	Altoona cake - Easterly	22.0	7.1	15.8	43.7	21.2	27.6	0.05	27.6	18.6	0.46
ACW	Altoona cake - Westerly	19.8	6.5	12.4	28.7	21.2	38.4	0.05	38.4	23.7	1.00
BC	Bellefonte cake	19.8	7.0	24.1	34.9	11.4	47.6	0.05	47.6	28.8	0.57
BPC	Baltimore Patapsco cake	22.6	5.5	11.5	15.2	14.2	74.0	13.3	60.8	21.9	1.63
BPP	Baltimore Patapsco pellets	95.7	5.6	10.6	12.5	12.3	63.0	1.27	61.7	17.6	0.60
BRC	Back River cake	23.4	7.8	15.1	19.9	72.5	54.3	9.18	45.1	27.2	0.21
BRP	Back River pellets	96.6	6.9	16.7	20.4	73.8	49.8	2.66	47.1	28.2	0.21
C	Control soil (Hagerstown)	98.7	5.2	27.5	1.93	24.3	2.1	0.59	1.54	0.64	0.01
DAP	DAP fertilizer	92.7	6.9	8.89	2.83	17.7	186	186	0.08	233	172
DC	Danville cake	12.8	7.6	7.77	23.1	11.8	60.6	4.53	56.1	14.4	0.19
DL	Danville LPT cake	18.7	12.2	5.33	100	7.79	39.8	2.89	36.9	10.4	0.44
DM	Dairy manure (PSU)	15.6	7.2	0.45	16.6	0.74	35.5	4.74	30.8	7.06	3.29
DOC	Dover cake	16.1	6.9	17.3	31.2	25.4	57.5	0.06	57.5	29.2	0.86
EC	Ephrata cake (Plant #2)	27.8	8.0	25.4	49.4	43.5	35.1	7.55	27.5	34.0	1.28
FPR	FPR (egg/omlett products)	3.9	6.4	4.23	39.9	4.49	58.3	11.3	46.9	34.6	10.01
LAN	Lancaster cake	23.6	6.2	29.5	15.0	24.8	52.1	6.08	46.0	20.5	0.84
LP	Largo pellets	94.6	6.0	5.85	22.6	10.4	73.5	0.30	73.2	35.0	4.05
MC	Madison cake	7.01	7.9	7.70	46.3	21.8	68.0	15.1	52.9	38.1	5.39
MC+WTR[(1)]	Madison cake + WTR	11.5	NM	23.2	28.0	13.6	40.3	8.62	31.6	22.5	1.59
NC	Nazareth cake	11.1	6.3	7.76	25.4	8.35	65.9	0.09	65.9	27.4	3.03
OC	Oceangro cake	15.2	7.9	18.3	31.6	31.9	62.4	9.38	53.1	33.8	0.57
OP	Oceangro pellets	95.4	7.0	17.5	30.4	30.1	56.7	2.00	54.7	32.1	0.36
PC	Philadelphia cake	32.7	8.2	21.6	22.5	91.5	38.1	10.7	27.4	25.8	0.30
PM1	Poultry manure #1 - layer	76.4	8.6	0.59	243	1.53	25.9	3.83	22.0	24.3	3.81
PM2	Poultry manure #2 - broiler	78.9	7.8	0.37	27.7	0.80	38.6	3.56	35.0	16.8	3.07
PM3	Poultry manure #3 - broiler	63.0	8.0	0.17	36.9	0.64	41.5	10.7	30.7	21.7	4.25
SM	Swine manure	5.1	6.5	2.25	22.7	3.13	65.4	51.0	14.5	26.4	10.32
SM+WTR[(1)]	Swine manure + WTR	7.54	NM	16.3	16.3	2.96	44.4	33.6	10.7	18.1	5.67
TSC	Tarpon Springs cake	15.4	7.5	18.6	35.1	29.6	65.6	3.83	61.8	38.3	0.95
TSP	TSP fertilizer	97.5	2.2	11.2	171	9.53	1.2	0.01	1.18	207	135
UAC	Upper Allen cake	12.7	7.0	20.7	35.8	3.38	62.9	0.06	62.9	36.6	1.09
UAJ	UAJA primary solids cake	29.1	7.0	13.1	19.5	2.92	33.2	3.19	30.0	12.6	0.91
UC	University Plant cake	18.1	7.0	5.65	43.3	52.7	52.1	3.70	48.4	35.1	0.14
WTR	Al water treatment residual	81.5	7.0	43.7	3.82	2.63	3.52	0.01	3.52	1.85	0.01
YC	York cake	16.3	8.0	8.50	42.1	10.1	68.8	10.1	58.7	36.1	2.37

(1) Elemental composition calculated from individual components. FPR = Food processing residual;
LPT = Lime post treatment; NM = Not Measured; WTR = Water treatment residual

WEP = Water-extractable P; PSU = Pennsylvania State University; FPR = food processing residual.

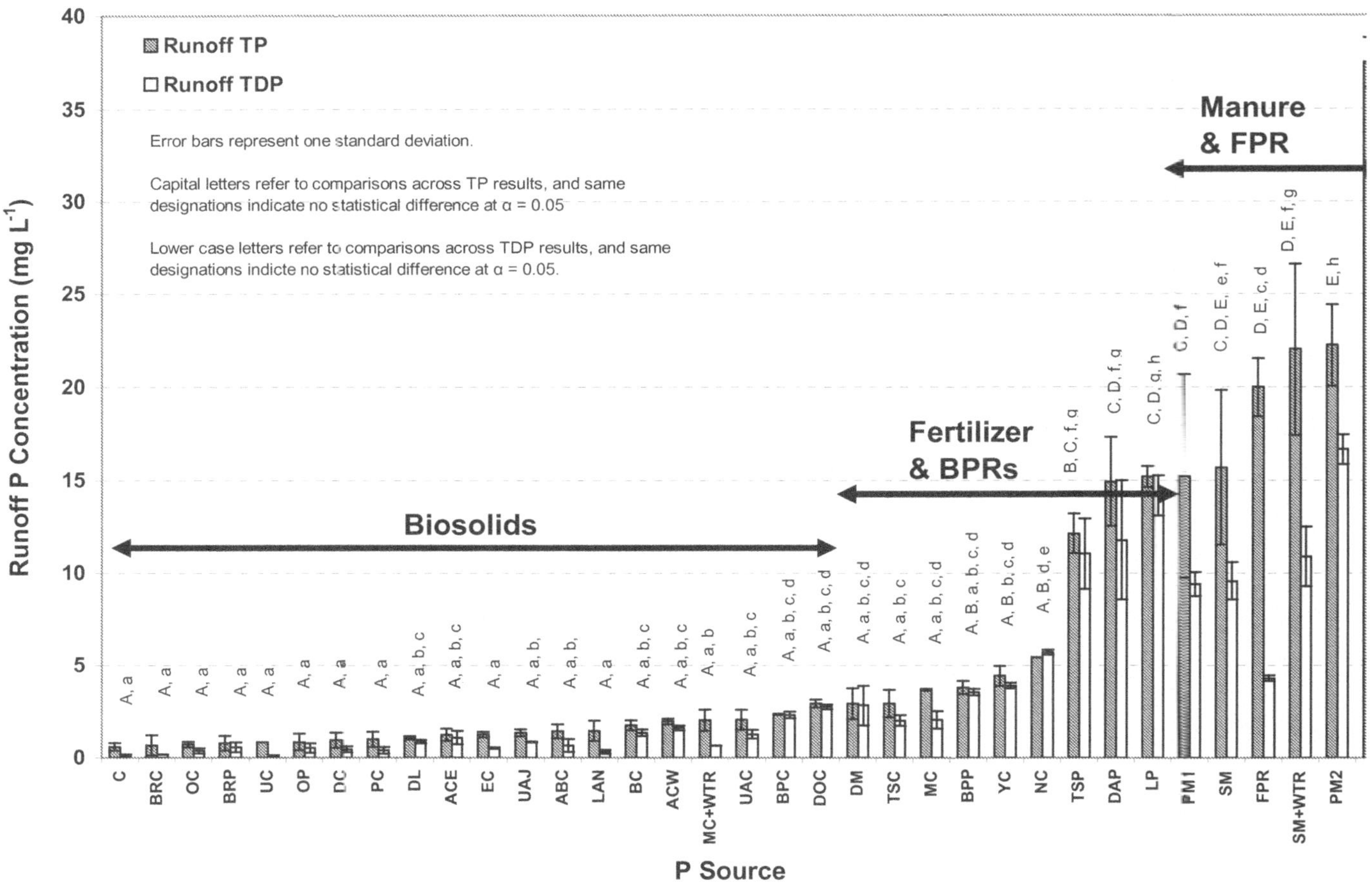

Figure 3-4. Runoff P Concentrations for 34 Materials, Each Applied at 100 kg P ha-1. Abbreviations as in Table 3-2.

Addition of WTR to the Madison BPR biosolids reduced total dissolved P losses from 2.05 mg L^{-1} to 0.66 mg L^{-1}, but statistically the total dissolved P levels are not different. The total dissolved P losses from the swine manure and from WTR-amended swine manure were also not statistically distinguishable. In fact, the total P and total dissolved P values suggest the WTR actually *increased* runoff losses from the swine manure. This unexpected result was likely caused by the short contact time between the WTR and swine manure, as the WTR was mixed with the P sources immediately before loading of the soil trays. It is believed that greater reaction time would have substantially reduced runoff P losses from the amended swine manure. Table 3-2 shows that addition of the WTR reduced the swine manure water-extractable P from 10.32 to 5.67 g kg^{-1}.

Despite extremely high water-extractable P values (152 and 185 mg kg^{-1} for DAP and TSP, respectively), the inorganic fertilizers resulted in intermediate total P and total dissolved P values. This has been observed for DAP by other researchers (Kleinman et al., 2005). They attribute this effect to the small soil area covered by the applied fertilizer, which results in substantial sorption of dissolved P in runoff in areas where the fertilizer was not applied. Alternatively, for fertilizer granules, complete dissolution likely did not occur during the relatively short duration of the rainfall experiment.

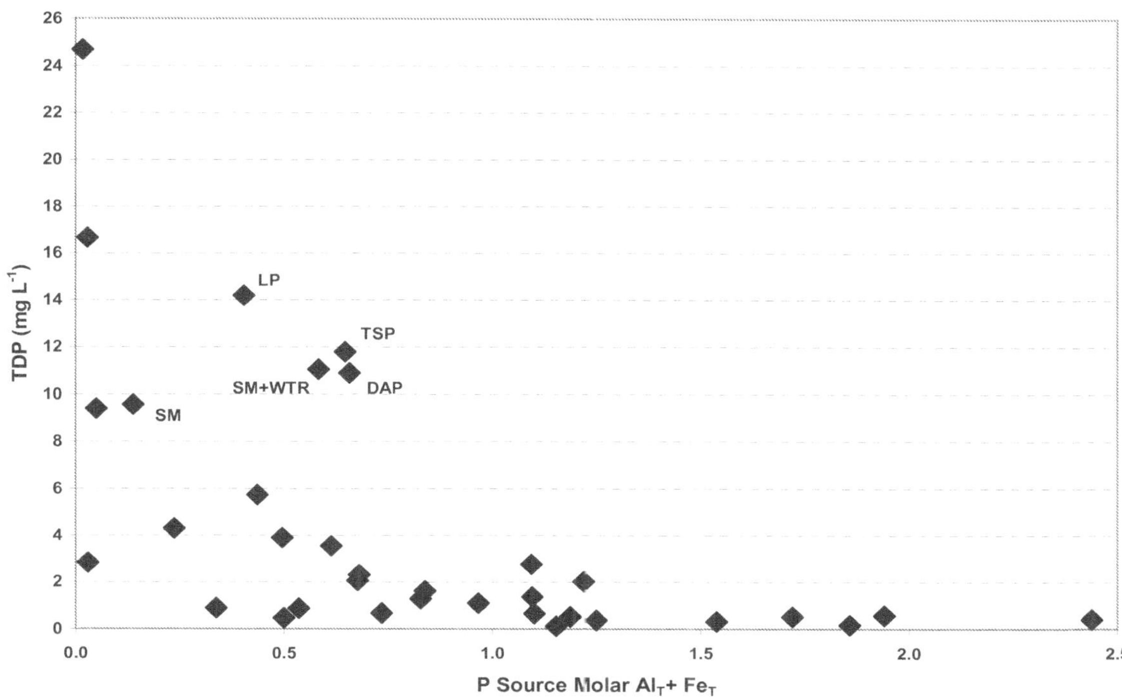

Figure 3-5. Total Dissolved P Concentration versus P Source Molar Al_T + Fe_T. LP, Largo pellets; SM, swine manure.

Several materials (Largo pellets, TSP, DAP, WTR-amended swine manure) appear to have exceptionally high total dissolved P given their Al + Fe content (Figure 3-5). Two of these are the fertilizer materials (TSP and DAP), which might be expected to behave differently than the organic sources of P because of the fertilizers' extremely high water-extractable P content. The Largo pellet biosolids is also substantially above the trend line. This material has repeated shown exceptionally high P loss behavior, even when compared to other BPR-generated biosolids. The relatively large total dissolved P of WTR-amended swine manure is believed to reflect the short contact time between the WTR and the manure.

Figure 3-6 shows the total dissolved P and total P values as a function of the P source total solids content. The data seem to fall into two distinct groupings. For materials with solids contents below 40%, which includes all biosolids except heat-dried products, greater solids content was associated with lower runoff P. This probably reflects the progressive removal of highly soluble P from the biosolids during solids dewatering. The second grouping (total solids > 60%) is comprised of three poultry manures, the two commercial fertilizers and the Largo pellets. The release of P from these materials therefore must be due to the highly soluble nature of the P source itself rather than from release of entrained water containing dissolved P. Three heat-dried materials, which had low runoff P (Oceangro pellets, Baltimore Patapsco pellets, Back River pellets), potentially might behave mechanistically more like materials in the first group.

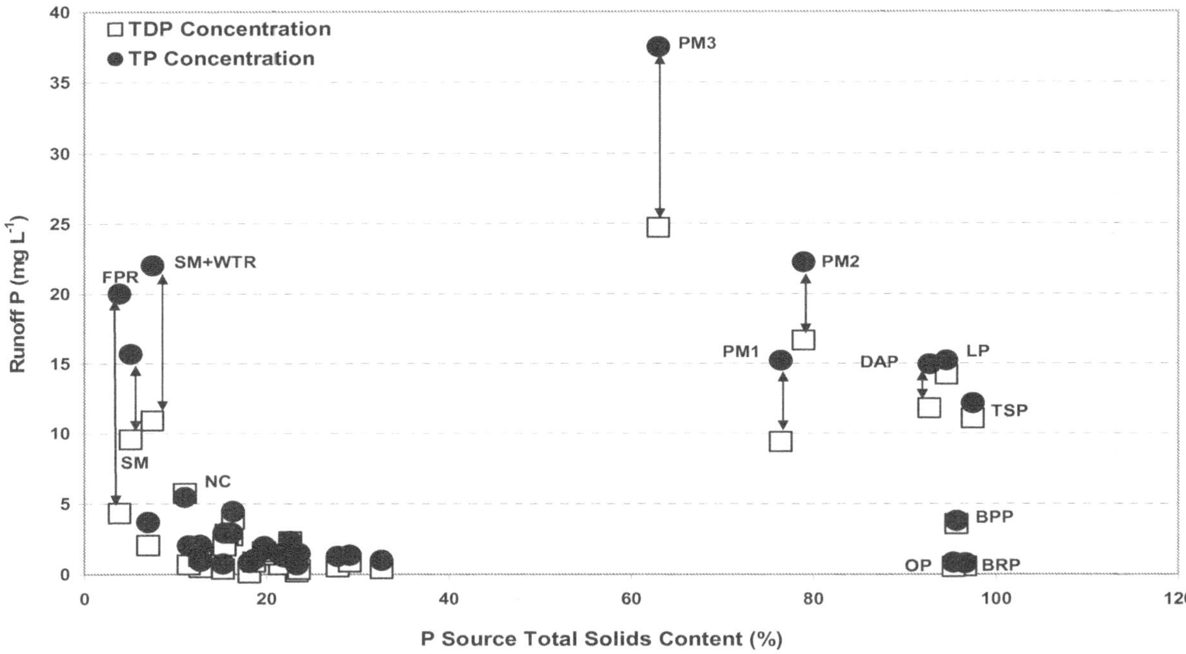

Figure 3-6. Total Dissolved P and Total P Concentration versus P Source Total Solids. See Table 3-2 for abbreviations.

3.3.2 Variable P Loading Rainfall Simulation

3.3.2.1 Soil and P sources

General properties and elemental levels for the unamended Hagerstown soil are shown in Table 3-3. The Mehlich-3 P concentration (95 mg kg^{-1}) exceeded typical crop nutritional requirements (Beegle, 2002), but fresh P source additions tend to mask impacts initial STP levels have on runoff dissolved P (Andraski and Bundy, 2003). The Hagerstown is typical of arable soils in Pennsylvania, and has been used in other P runoff studies (McDowell and Sharpley, 2003).

Ten biosolids from eight WWTPs were chosen to represent the range of materials typically applied to agricultural soils (Table 3-4). Two biosolids, Largo pellets and Danville limed cake, were also selected to assess the effects of heat drying and lime stabilization on runoff P loss. Three biosolids cake materials (Bellefonte, Danville, and University) were generated by conventional aerobic or anaerobic digestion stabilization. The Largo facility employs BPR where the P, removed from wastewater as part of the microbial biomass, is highly water soluble (Brandt et al., 2004). Three materials were used as heat-dried pellets (Largo, Patapsco, Back River), and the Syracuse biosolids was processed by advanced alkaline stabilization (the N-Viro® process).

All P sources were applied to the soil trays at a rate equivalent to134 kg plant-available N ha^{-1}, a typical recommendation for corn in Pennsylvania. For the biosolids, agronomic N availability was based on state regulatory prescriptions for ammonia availability and organic N mineralization. Dairy manure N availability was based on recommendations provided in Beegle (2002). The resulting P application rates ranged from 122 (dairy manure) to 555 (Syracuse N-Viro® biosolids) kg P ha^{-1} (Table 3-4).

Table 3-3. Properties of Unamended Hagerstown Soil.

Parameter	Value
pH	6.2
Acidity, cmol kg^{-1}	2.0
CEC, cmol kg^{-1}	12.6
WEP, mg kg^{-1}	0.02
Elemental Levels	
Al, g kg^{-1}	32.9
Fe, g kg^{-1}	26.2
Ca, g kg^{-1}	2.2
Mg, g kg^{-1}	3.6
P, g kg^{-1}	0.90
Mehlich-3 Extractable	
P, mg kg^{-1}	95
K, mg kg^{-1}	299
Mg, mg kg^{-1}	168
Ca, mg kg^{-1}	1694

The dairy manure was also treated with alum or $FeCl_3$ solutions, designated as DMA and DMF, respectively. Just prior to application of the manure to the soil boxes, the salt solutions were mixed into the manure such that the combined molar concentrations of the added Al and Fe were equal to the total molar P in the dairy manure (0.18 mol P kg^{-1}).

Although nutrient content, physical properties and chemical constituents differ because of various wastewater and stabilization processes (Table 3-4), the biosolids used in this study are typical of those produced nationally and applied to agricultural land (U.S. EPA, 1995). The solids content ranged from dewatered moist "cake" materials (13–27%) to heat dried products (>90%). Mean total P content for the biosolids was 23 g kg^{-1}, in close agreement with median P of 22 g kg^{-1} reported by Stehouwer et al. (2000) for all biosolids submitted during 1997 to the Penn State Agricultural Analytical Services Laboratory. Lower concentrations of P in Danville limed cake and Syracuse N-Viro reflect dilution from addition of lime and alkaline by-products. The water-extractable P levels in the biosolids were typically below 1 g kg^{-1}, with the BPR-produced Largo materials being significantly higher. The Al and Fe contents were variable, reflecting additions of Al- and Fe-salts in wastewater treatment and solids processing or, in the case of Philadelphia cake, the discharge of Fe-based WTRs to sanitary sewers. The total P and water-extractable P of the dairy manure fall within the range of values previously reported (Sharpley and Moyer, 2000; Brandt et al., 2004).

3.3.2.2 P Source Effects on Runoff Phosphorus

The effects of P sources on total dissolved P and total P concentrations in runoff followed a similar trend for the first (Figure 3-7a) and second rainfall events (Figure 3-7b). The P concentrations in runoff were generally greater in the first event than the second event. Other studies have documented that total and dissolved P in runoff decreased with successive rainfall events following surface application of livestock manures (Sharpley, 1997) and biosolids (Rostagno and Sosebee, 2001; Penn and Sims, 2002). For biosolids-amended field plots, Sims et al. (2003) found the initial event accounted for 52–73% of the total runoff P collected over four rainfall events.

Table 3-4. P Source Characteristics and Loadings Used in the Variable P Loading Rainfall Simulation.

P-Source	Label	Description	WEP (g kg^{-1})	Solids (%)	Total (g kg^{-1})					Dry Matter Application Rate (kg tray^{-1})	P Application Rate (kg ha^{-1})	PSC[1]
					Al	Fe	Ca	N	P			
University cake	UC	Conventional treatment without Al/Fe addition	0.36	12.2	5.5	9.1	32.8	70.9	17.6	0.175	154	0.3
Danville cake	DC	Anaerobic digestion	0.38	12.7	11.1	15.1	25.6	65.8	19.8	0.183	181	0.3
Largo pellets	LP	Heat dried BPR, Class A	4.55	94.6	5.9	10.4	22.6	73.5	34.9	0.108	187	0.8
Danville limed cake	DL	Lime stabilized	0.20	26.8	5.0	5.4	115.0	26.8	8.6	0.449	193	0.4
Patapsco (Baltimore) pellets	PB	Heat treated, low Fe	0.84	95.0	7.5	12.5	13.3	58.1	17.2	0.228	196	0.2
Largo cake		BPR wet-side processing	6.76	16.9	5.5	9.2	22.6	78.2	33.4	0.127	213	0.8
Bellefonte cake	BC	Aerobic digestion, high Al	1.31	15.3	25.9	7.3	20.6	63.8	31.2	0.141	220	0.3
Back River (Baltimore)	BR	Heat-treated, high Fe	0.26	90.1	15.1	53.5	31.4	53.7	29.4	0.222	326	0.2
Philadelphia cake	PC	Anaerobic digestion, high Fe	0.19	26.3	14.3	72.3	18.7	36.9	26.4	0.280	369	0.3
Syracuse N-Viro	SN	Advanced alkaline stabilization	0.01	60.5	8.2	21.9	260	7.2	6.79	1.63	555	0.1
Dairy manure	DM	Fresh dairy manure	3.41	13.0	1.3	1.5	22.3	33.1	5.7	0.428	122	0.8
Dairy manure + Fe	DMF	$FeCl_3$-treated dairy manure	1.07	13.0	1.3	11.8	22.3	33.1	5.7	0.428	122	0.5
Dairy manure + Al	DMA	Alum-treated dairy manure	0.31	13.0	6.3	1.5	22.3	33.1	5.7	0.428	122	0.5

(1) P source coefficient (PSC) from Pennsylvania P Index. PSC value of 0.1 for Syracuse N-Viro, though not currently in index, has been proposed for P-sources with extremely low W. WEP = water-extractable P.

(a)

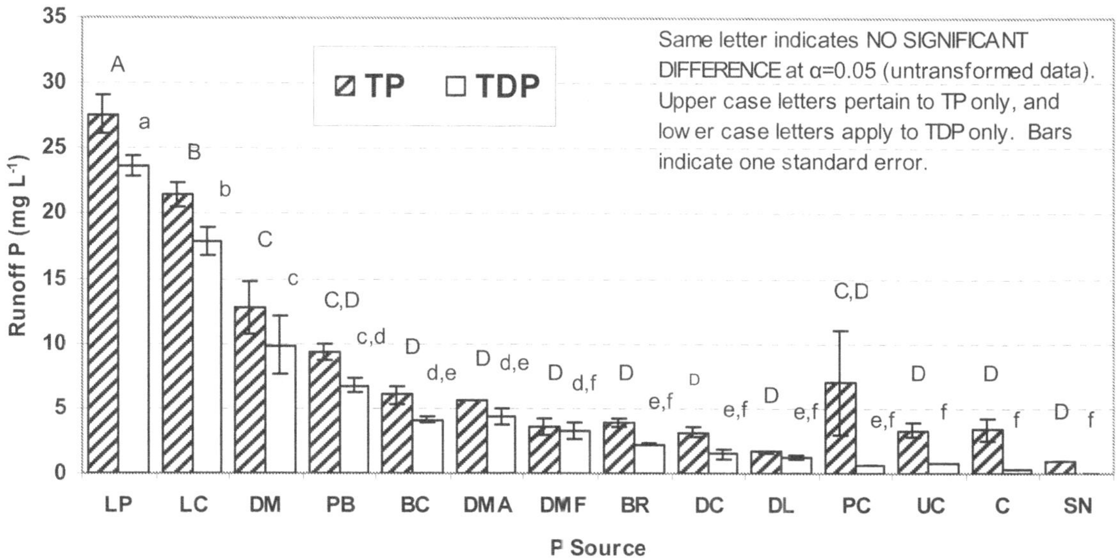

(b)

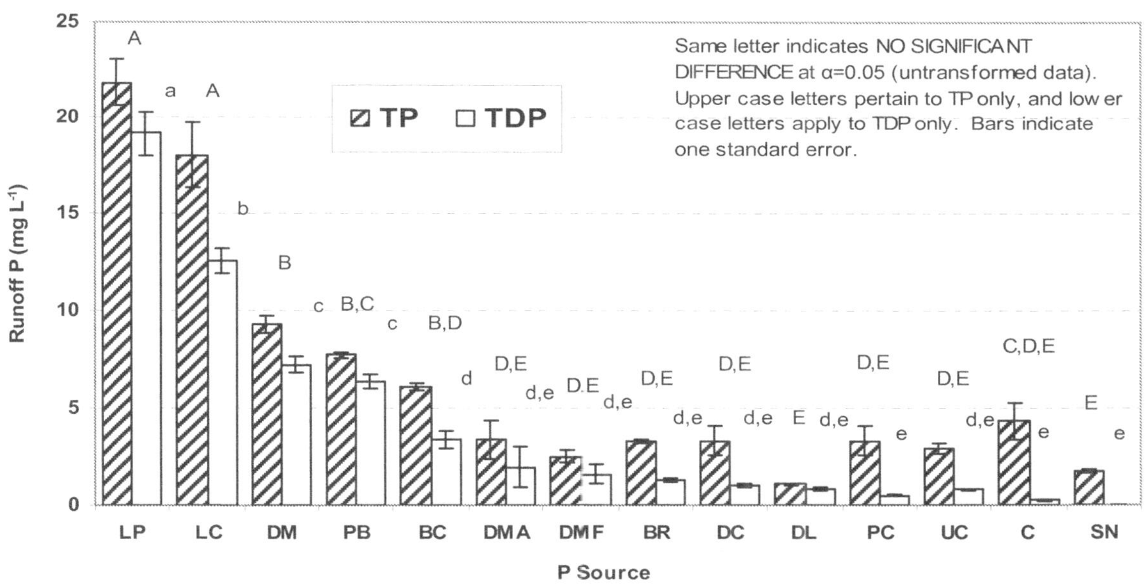

Figure 3-7. Runoff P for Various Sources Applied at N-Based Rates: (a) First Runoff Event, (b) Second Runoff Event. See Table 3-4 for abbreviations. TP, total P; TDP, total dissolved P.

Most of the runoff P in the unamended control treatments was in the particulate form with <10% as total dissolved P. However, where P sources were surface applied, the proportions of total dissolved P in runoff increased to 12–84% of total P. Because the P sources had variable N and solids contents, trays received different dry matter application rates (Table 3-4). Sources that effectively covered the tray surface (e.g., dairy manure) had particulate P lower than that of the control treatments, likely due to a mulching effect (sealing) of the soil surface. For such P sources, the particulate P was likely derived primarily from the P source rather than from the soil. For three P sources with the lowest mean total dissolved P levels (Philadelphia cake, University cake, and Syracuse N-Viro), the proportion of particulate P to total dissolved P was much greater than P sources with high total dissolved P concentrations. Withers et al. (2001) similarly found that surface-applied P sources with the lowest total P losses had higher proportions of particulate P than dissolved P loss.

Although the dairy manure had the lowest total P rate (122 kg P ha^{-1}), it had greater total dissolved P than all biosolids except the Largo cake and pellets, and the Baltimore Patapsco pellets. The generally greater runoff P concentration for dairy manure compared to biosolids was shown previously where the P sources were all applied at 100 kg P ha^{-1} (Brandt and Elliott, 2003). In this study, the lower water solubility of the P in most biosolids offset the higher total P application rates relative to the dairy manure. The total dissolved P of the Patapsco (Baltimore) pellet treatment was statistically similar to the dairy manure treatment, despite a 61% greater total P application rate for PB. Notably, the total P and total dissolved P for six of the biosolids treatments (University cake, Danville limed cake, Back River, Philadelphia cake, Syracuse N-Viro) applied at 154–555 kg P ha^{-1} were not statistically different from the control soil treatment (Figures 3-7). Elsewhere, Withers et al. (2001) reported that P in runoff from biosolids-treated field plots was not significantly different from the control plot.

The Largo BPR materials (Largo pellets and cake) produced statistically greater total dissolved P and total P than the dairy manure treatment (Figure 3-7). As a category, BPRs have the greatest water-extractable P values among biosolids types (Brandt et al., 2004). The water-extractable P values for the Largo pellets and cake biosolids were 4.55 and 6.76 g kg^{-1}, respectively. The greater water-extractable P values of the Largo cake and Largo pellets coupled with the higher total P application rates to meet the common plant-available- N requirement (Table 3-4) resulted in greater total dissolved P compared to the dairy manure (water-extractable P = 3.41 g kg^{-1}). However, the low water-extractable P for most biosolids (water-extractable P < ~1 g kg^{-1}) suggests lower P runoff potential compared to typical livestock manures, assuming biosolids and dairy manure have no effect on runoff amount. Penn and Sims (2002) also found that biosolids produced by treatment of wastewater using BPR exhibited the greatest total dissolved P in runoff from rainfall simulation experiments.

The Danville cake and limed biosolids exhibited total dissolved P and total P concentrations that were not statistically different (Figure 3-7) where P application rates were nearly identical (Table 3-2). Other studies have generally found that P solubility is greater in soils amended with limed biosolids, at least relative to Fe- or Al-dominated biosolids systems (Penn and Sims, 2002). Jokinen (1990) assessed the influence of treatment process on extractable soil P in biosolids amended soils and concluded that, while Al-sludges tended to reduce P solubility, Ca-sludge increased soluble P values in soils. Soon and Bates (1982) reported that the extractability of P from soils amended with chemically treated, anaerobically digested sludges followed the order: Ca-sludge >> Fe-sludge ≥ Al-sludges. The presence of lime in biosolids treated with metal salts tended to result in higher water-soluble P in the amended soils (Maguire et al., 2000). Leytem et al. (2004) suggested that the higher solubility was caused by the release of Ca–P complexes in response to the acidity of the amended soils. Such a mechanism would not apply to this study because the materials were not incorporated into the soil. In this case, the similar runoff P concentrations from the limed and unlimed Danville biosolids (Figure 3-7) reflect the nearly equal water-extractable P values (Table 3-4).

The Syracuse N-Viro had the greatest total P application rate (555 kg P ha^{-1}), but resulted in negligible soluble P in runoff (Figure 3-7a). This was a reflection of the extremely low water-extractable P concentration (0.01 g kg^{-1}) of the N-Viro material (Table 3-4). Richards et al. (1997) attributed the low extractability and mobility of P in the Syracuse N-Viro product to P precipitation reactions occurring at elevated pH and Ca levels. Advanced alkaline stabilization (e.g., the N-Viro process), involves mixing biosolids with large quantities of alkaline materials so that the mixture is dried and pasteurized though heat generation. Although not statistically different ($\alpha = 0.05$), the mean total dissolved P and total P concentrations of the Syracuse N-Viro treatment were lower than the control treatment. Maguire et al. (2000) suggested that some biosolids increase the P sorption capacity of the soils and thus reduce P into runoff and drainage waters.

Treatment of the dairy manure with Al or Fe reduced P concentrations in runoff by about one-half compared to the unamended manure (Figure 3-7). This reduction likely reflected the lower water-extractable P concentrations of the amended dairy manure (Table 3-4). Alum additions to poultry manure have been shown to reduce runoff P (Moore et al., 1999). Here, the research team demonstrates that total dissolved P concentrations in runoff can potentially be reduced by adding Al and Fe salts to materials such as dairy manure that have high concentrations of water-soluble P.

The importance of P source Al and Fe content is further illustrated in Figure 3-8, which shows that the total dissolved P levels for the first runoff event were consistently lower when the total molar Al + Fe content of the P sources was high. The critical role of Al and Fe in determining P solubility and release from biosolids amended soils has been documented previously (Maguire et al., 2001; Brandt and Elliott, 2003). Note that several materials with molar Al + Fe contents about 0.3 to 0.4 had widely varying total dissolved P concentrations. Although very high Al + Fe contents may result in low runoff total dissolved P, other factors (P loading rate, wastewater treatment, and solids processing) also influence susceptibility to P loss.

3.3.2.3 P Loading Rate Effects on Runoff Phosphorus

Runoff total dissolved P and total P concentrations generally increase with P application rate when manures are surface broadcast. Edwards and Daniel (1993) reported dissolved reactive P levels in runoff of 0.8, 11.9, and 29.4 mg L^{-1} from grassed plots broadcast with swine manure at rates of 0, 19, and 38 kg total P ha^{-1}, respectively. Rostagno and Sosebee (2001) found that dissolved reactive P and total dissolved P in runoff water increased with biosolids application rates (0 to 90 Mg ha^{-1}). Kleinman and Sharpley (2003) reported that the concentrations of dissolved reactive P and total P in runoff increased as the total P application rate from broadcast manures increased from 0 to 150 kg P ha^{-1}. They attributed this to an increase of soluble manure P to runoff water as more manure was applied, because dissolved reactive P became a higher proportion of runoff total P as application rate increased. The effect was particularly pronounced for total P application rates > 75 kg ha^{-1}.

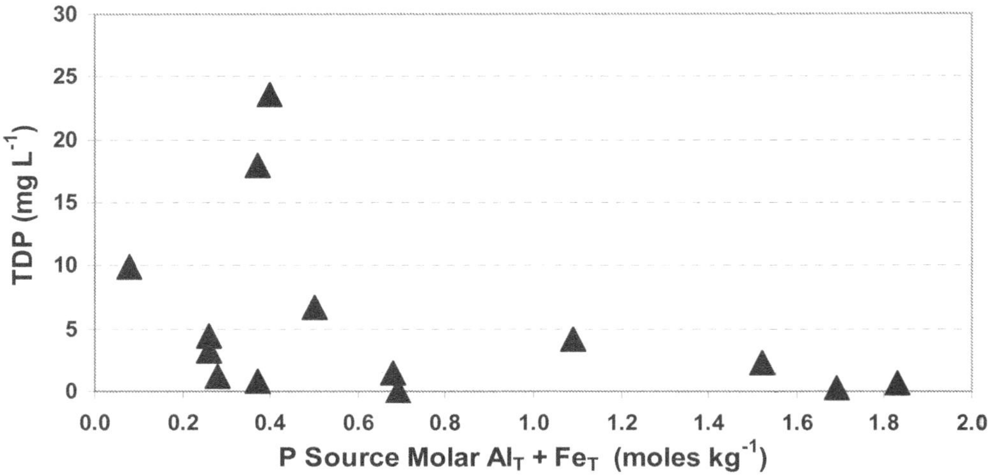

Figure 3-8. Runoff P as a Function of Source Total Fe + Al Concentration.

A positive correlation between runoff dissolved P and application rate (Kleinman and Sharpley, 2003) is logical for a given P source type since the amount of soluble P available to runoff water depends on the application rate. Such a relationship should not necessarily hold for a broad range of P sources where the potential to release P into runoff water is highly variable. Figure 3-9 shows the total dissolved P and total P concentrations, respectively, in runoff for rainfall event 1 for all 14 treatments (10 biosolids, three manures, and one control) as a function of the total P application rate (kg ha^{-1}). Clearly, there is no positive correlation between runoff P and the total P application rate. The Syracuse N-Viro treatment had the highest total P application rate (555 kg ha^{-1}), but the lowest total dissolved P concentrations in runoff (<0.1 mg L^{-1}). There were seven biosolids types with similar total P applications rates (~180–220 kg ha^{-1}), but the mean runoff total dissolved P values were from <1 mg L^{-1} to >20 mg L^{-1} (Figure 3-9a). Others have also shown that biosolids applied at the same total P rate exhibit significantly different runoff P losses (Penn and Sims, 2002; Brandt and Elliott, 2003; Sims et al., 2003).

Runoff P concentrations were related to the water-extractable P levels of the various P sources, however (Figure 3-10a,b). For the wide range of materials in this study, water-extractable P and total dissolved P were linearly related ($r^2 = 0.83$) for the first rainfall event (Figure 3-10a). Because total dissolved P comprised most of the total P (Figure 3-7a), a similarly strong relationship ($r^2 = 0.81$) was found between water-extractable P and total P in runoff (Figure 3-10b). Kleinman et al. (2002a) observed that dissolved reactive P losses from soils surface amended with dairy, poultry and swine manures (all at 100 kg P ha^{-1}) were proportional to the water-extractable P of the applied P sources. Withers et al. (2001) reported that differences in runoff P from TSP fertilizer, cattle slurry manure, and two biosolids were more related to P-source water-extractable P than total P application rates, although the latter was 150–329 kg P ha^{-1}.

(a)

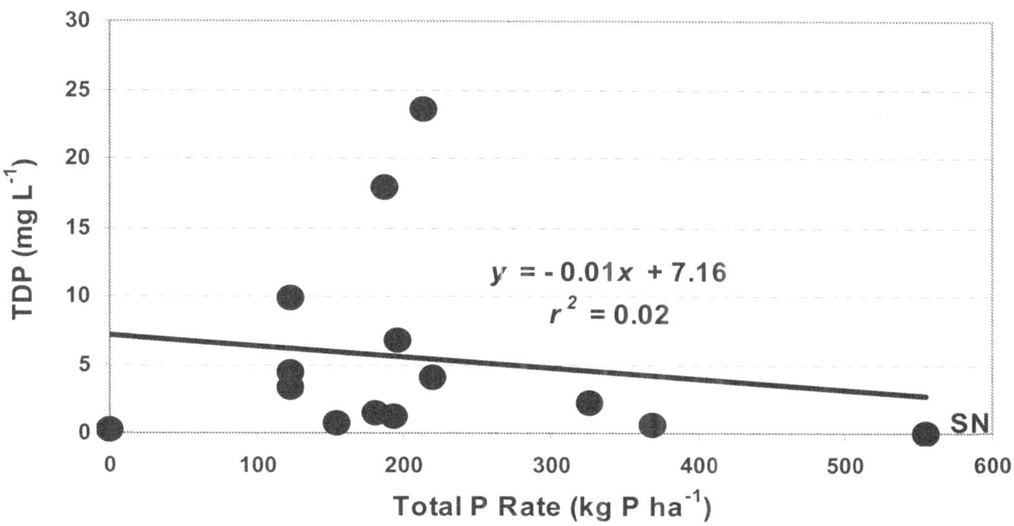

(b)

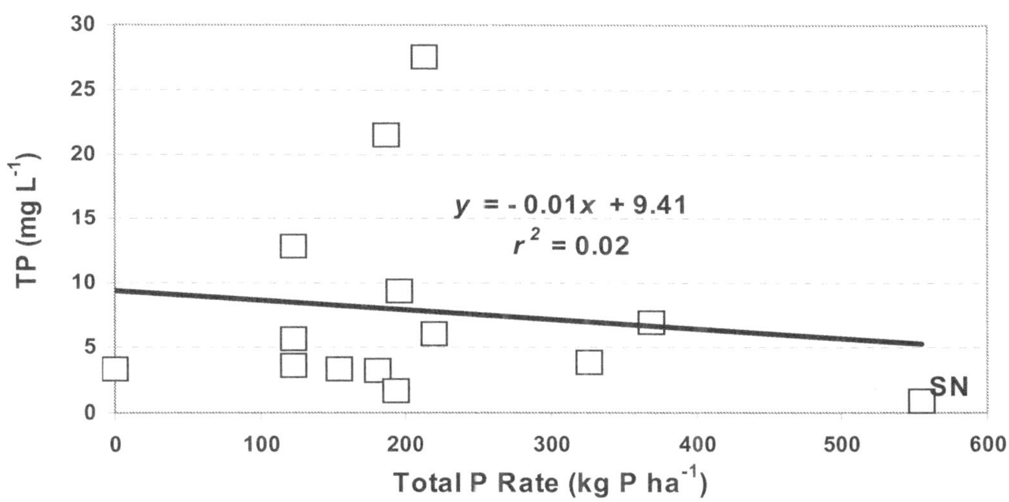

Figure 3-9. Runoff P (a) Total Dissolved P, and (b) Total P as a Function of Total P Application Rate.

Some states (Pennsylvania, Arkansas) use P-source water-extractable P concentration as an indicator of P loss potential in site assessment indices (Weld et al., 2000). In Pennsylvania, the water-extractable P values have been used to develop PSCs as quantitative indicators of the relative availability of the P in land-applied materials to be transported in runoff and drainage. These PSCs are fractional values referenced to mineral fertilizer (i.e., fertilizer PSC = 1.0). In the Pennsylvania P index, PSCs are multiplied by the organic P source application rate to obtain the fraction of total applied P available for transport. The PSCs for the materials in this study are given in Table 3-4. When the total P application rate is multiplied by the respective PSC, the materials should theoretically be normalized on a soluble-P application rate basis.

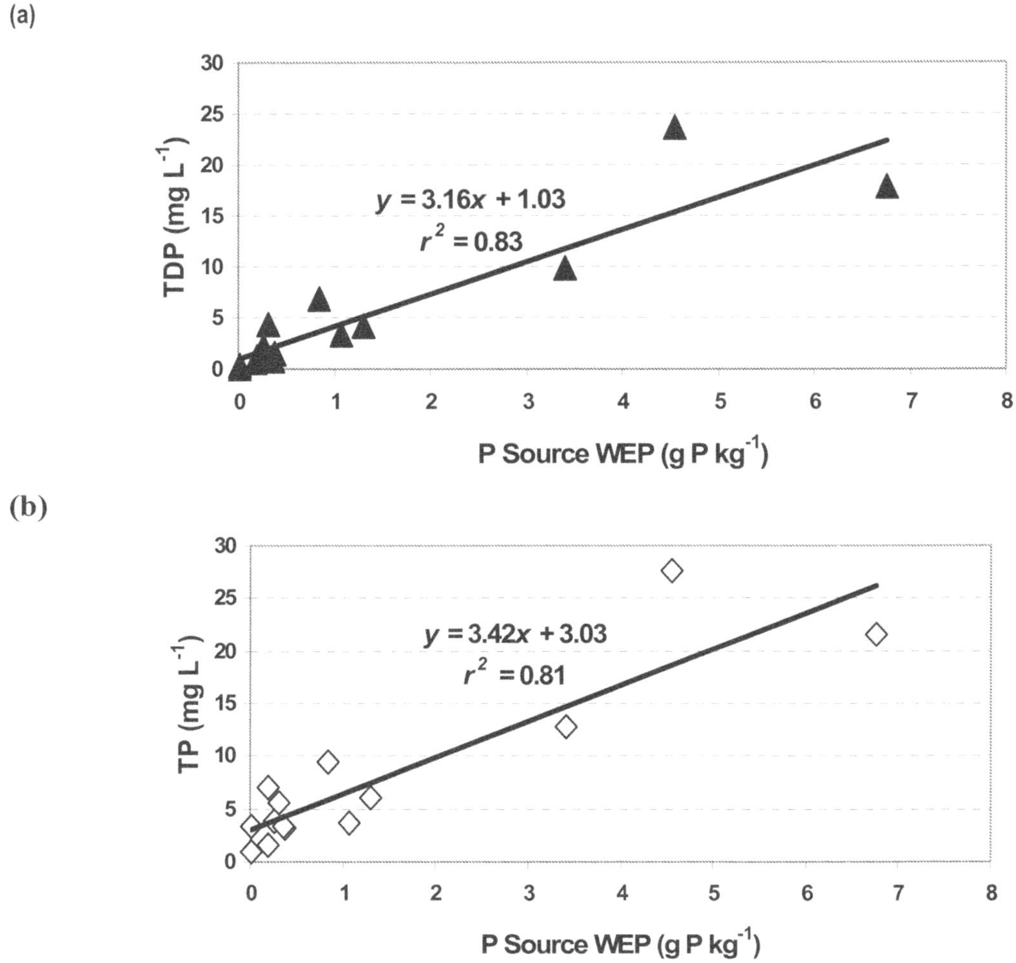

Figure 3-10. Runoff P (a) Total Dissolved P, and (b) Total P as a Function of Source Water-Extractable P (WEP).

Figure 3-11 show that total dissolved P and total P are strongly correlated with total P application rates for multiple sources when multiplied by the respective PSC values. Data clearly illustrate that runoff P losses from most biosolids (eight of 10) are less than from dairy manure, despite greater total P application rates for the biosolids. The BPR biosolids (Largo cake and Largo pellets) had greater total P application rates than the dairy manure (Table 3-4), but also exhibited high concentrations of total dissolved P and total P in runoff. In the current Pennsylvania P index, dairy manure and BPR biosolids both have PSC values of 0.8. The data point for Philadelphia cake does not fit well in the correlation. This material has exceptionally high Fe (72.3 g kg^{-1}) for biosolids because Fe-based WTRs are disposed into the sanitary sewers in Philadelphia. Such high Fe causes the water-extractable P to be very low (0.19 mg kg^{-1}), which translates into low potential for P loss. The PSC used in this correlation was 0.3 for conventionally treated biosolids because no category currently exists for "high Fe" biosolids in the PA P index. The results of Leytem et al. (2004) suggest that biosolids containing high levels of Fe should have lower PSC values than conventionally treated materials.

The correlations in Figure 3-11 are remarkably good considering that the PSC values are default values. Actual P sources tend to defy rigid categorization and, ideally, PSC values should be used that are unique to the specific P source under evaluation. Algorithms for converting laboratory-determined water-extractable P values into material-specific PSC values for use in P indices appear to hold promise (see Section 4.2.2.3). This approach addresses the problem of defining compositional thresholds for classifying a material as a "high Fe" biosolids, or the dilemma of selecting a table-value PSC for P sources that conceivably fit into more than a single category.

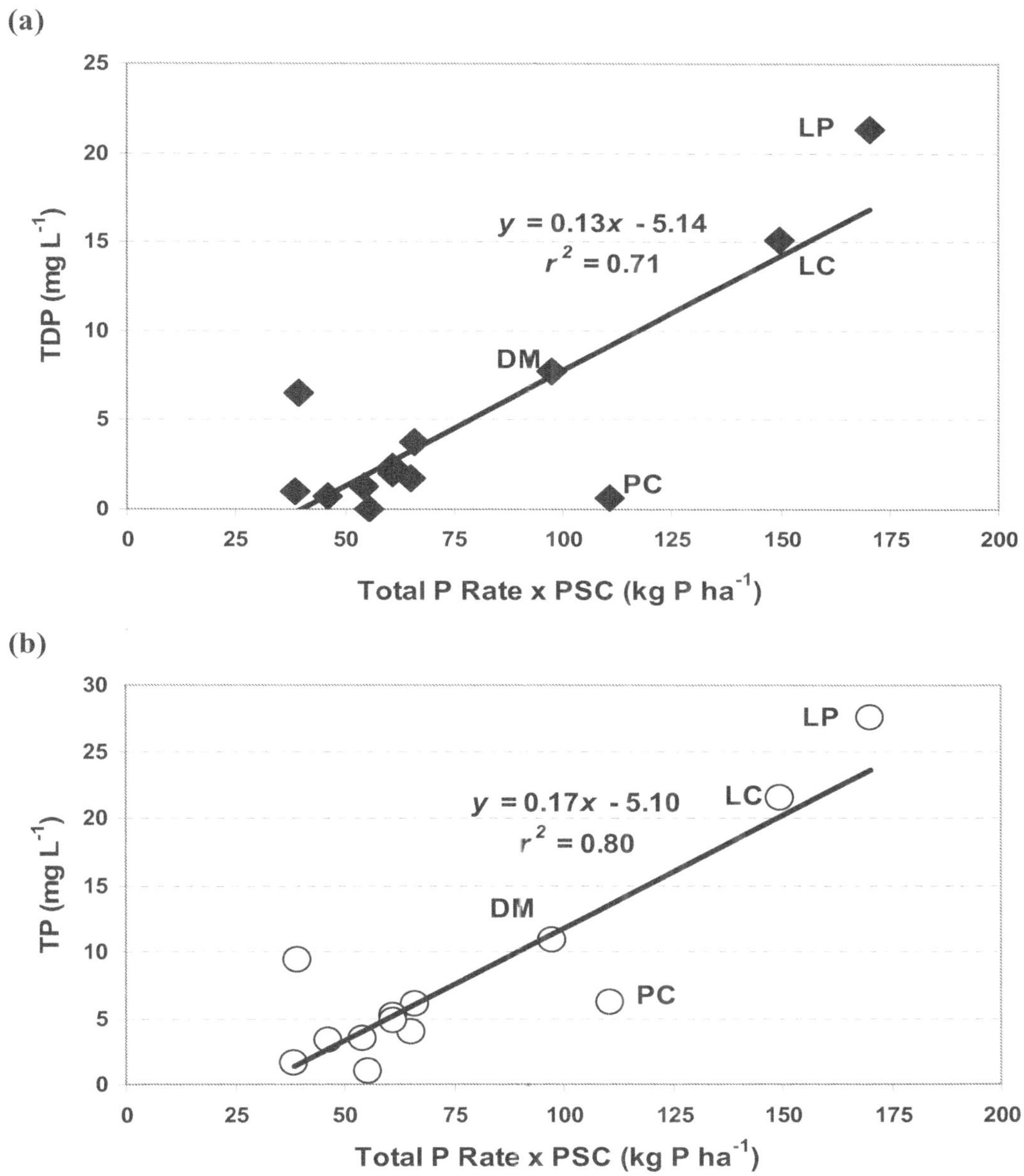

Figure 3-11. Runoff P (a) Total Dissolved P, and (b) Total P as a Function of Source Rate Corrected for P Source Coefficient. DM, dairy manure; LC, Largo cake; LP, Largo pellets; PC, Philadelphia cake.

3.3.2.4 Conclusions for Variable P Loading Study

There are significant differences in runoff P concentrations among biosolids types following surface application to soil boxes exposed to simulated rainfall. Experimental conditions mimic spreading of biosolids without incorporation. Penn and Sims (2002) found short-term differences in runoff P losses even when various biosolids types were mixed thoroughly with soil before simulated rainfall. As biosolids equilibrate with the soils over time, soil properties play an increasingly important role in the release of P to runoff. The P concentration in runoff from biosolids-amended soils depends on the types of wastewater and solids processing methods used to generate the biosolids. Some biosolids types (e.g., BPR products) resulted in higher runoff P concentrations than dairy manure applied at the same plant-available N rate. Other biosolids, because of additions of Al and/or Fe during wastewater treatment, or through solids processes like heat drying, produced runoff P losses not statistically different from unamended soil treatments. The data provide strong evidence that the solubility of the P in the organic amendment exerts a major influence on the potential for offsite migration of P at land application sites.

Findings of this and other (Leytem et al., 2004) studies confirm the need to incorporate weighting coefficients into P site indices to reflect the portion of total applied P susceptible to off-site transport. Of the 41 state P indices evaluated by Sharpley et al. (2003), only nine states (Arkansas, Delaware, Florida, Georgia, Louisiana, Maryland, Pennsylvania, Tennessee, and Virginia) include an estimate of the availability or solubility of the P applied to the site. Of these states, not all make specific reference to biosolids or, if they do, a single coefficient is used for all types of biosolids. If P index use is confined to one type of P source with relatively constant water-extractable P, inclusion of PSC in the index is less important.

For biosolids and treated manures, however, the susceptibility of P to leaching and runoff is so variable that the ability of a P index without PSCs to accurately predict site vulnerability risk is compromised. Moreover, unless the variable and typically lower solubility of P in biosolids is taken into account, the potential for P loss from biosolids application fields will likely be overestimated by site index assessment tools.

3.3.3 WTR Amendment Simulation

3.3.3.1 Soil and P Sources

This rainfall-runoff experiment employed the biosolids/manure/soil combinations evaluated in the SFWMD field study (see Section 3.2.1). Thus, fertilizer-P (TSP), biosolids from Boca Raton (high-soluble P) and Pompano Beach (low-soluble P) WWTPs, poultry (layers) manure, and an Al-WTRs from the Manatee County (Fla.) water treatment plant were evaluated. Detailed analyses of the P source materials and WTR are presented in Appendix Table A-2.

The P_T values are typical of manures and biosolids and the analysis confirmed that the TSP is ~20% total P. The WTR contains significant P_T, so that loading at 1% (22.4 Mg ha^{-1}) represents a substantial P loading (121 kg P ha^{-1}) to the runoff trays. The P sources were applied at 56 and 224 kg P ha^{-1}. The Immokalee A-horizon fine sand was used as the soil base. The Immokalee A is acidic, fine sand with negligible P-sorbing capacity (Appendix Table A-1).

3.3.3.2 Rainfall Simulation Results

Because P leaching was expected to be significant in the Immokalee fine sand, the national protocol for rainfall simulation boxes was altered to allow collection of runoff and leachate simultaneously (see Section 2.2.3).

The various forms of P in runoff and leachate (soluble inorganic, soluble organic, particulate, and total) were expressed as flow-weighted concentrations to normalize P loss in the 3 rainfall events. Others have emphasized that the majority of P loss occurs in the first rainfall event, but substantial P losses occur in subsequent events. Expressing the data as flow-weighted averages accounts for loses in all events, while allowing expression of the losses as concentrations.

Flow-weighted concentrations of various P forms in runoff are presented in Figure 3-12. Total dissolved P was mostly inorganic P (dissolved organic P <5% of total dissolved P). Total runoff P greatly exceeded total dissolved P (especially for the organic sources of P treatments), reflecting the significant contribution of particulate P to total runoff P. Total dissolved P values for the organic sources tracked well with the percentage of water-extractable P values of the sources, but not for TSP because most of the fertilizer-P loss was in leachate.

Flow-weighted concentrations of the various P forms in leachate (Figure 3-13) were dominated by dissolved inorganic P. There was minimal dissolved organic P in leachates and variable, but generally small, amounts of particulate P. Total dissolved P in leachate was much greater in leachates than in runoff for all treatments, but especially for fertilizer-P. The data emphasize the dominant role of leaching as a P loss mechanism in sands with low P-sorbing capacity. Such soils dominate much of the Coastal Plain region of the United States. Site assessment tools for these soils must include consideration of leaching, as well as runoff, to realistically access P loss risk.

Given the importance of particulate P, especially in runoff for organic sources of P, consideration of the potential particulate P lability seems appropriate. Sharpley (1993a,b) showed that P extracted from sediment-ladened runoff by ISP was strongly correlated with the growth of P-starved *Selenastum capicornutum*, and suggested that ISP could be used as a good estimate of biologically available P (BAP). The BAP should include dissolved P forms and the labile portion of particulate P. Summing the BAP masses for runoff and leachate should access the total pool of P lost that is of environmental significance.

Flow-weighted BAP concentrations for runoff and leachate are presented in Figure 3-14 a and b, respectively. BAP concentrations are generally greater in leachate than runoff, especially for fertilizer-P, and concentrations in leachate track well with water-extractable P values: TSP (water-extractable P = 175 g kg^{-1}) >> poultry manure (water-extractable P = 4.6 g kg^{-1}) ≈ Boca Raton biosolids (water-extractable P = 5.5 g kg^{-1}) > Pompano biosolids (water-extractable P = 1.2 g kg^{-1}).

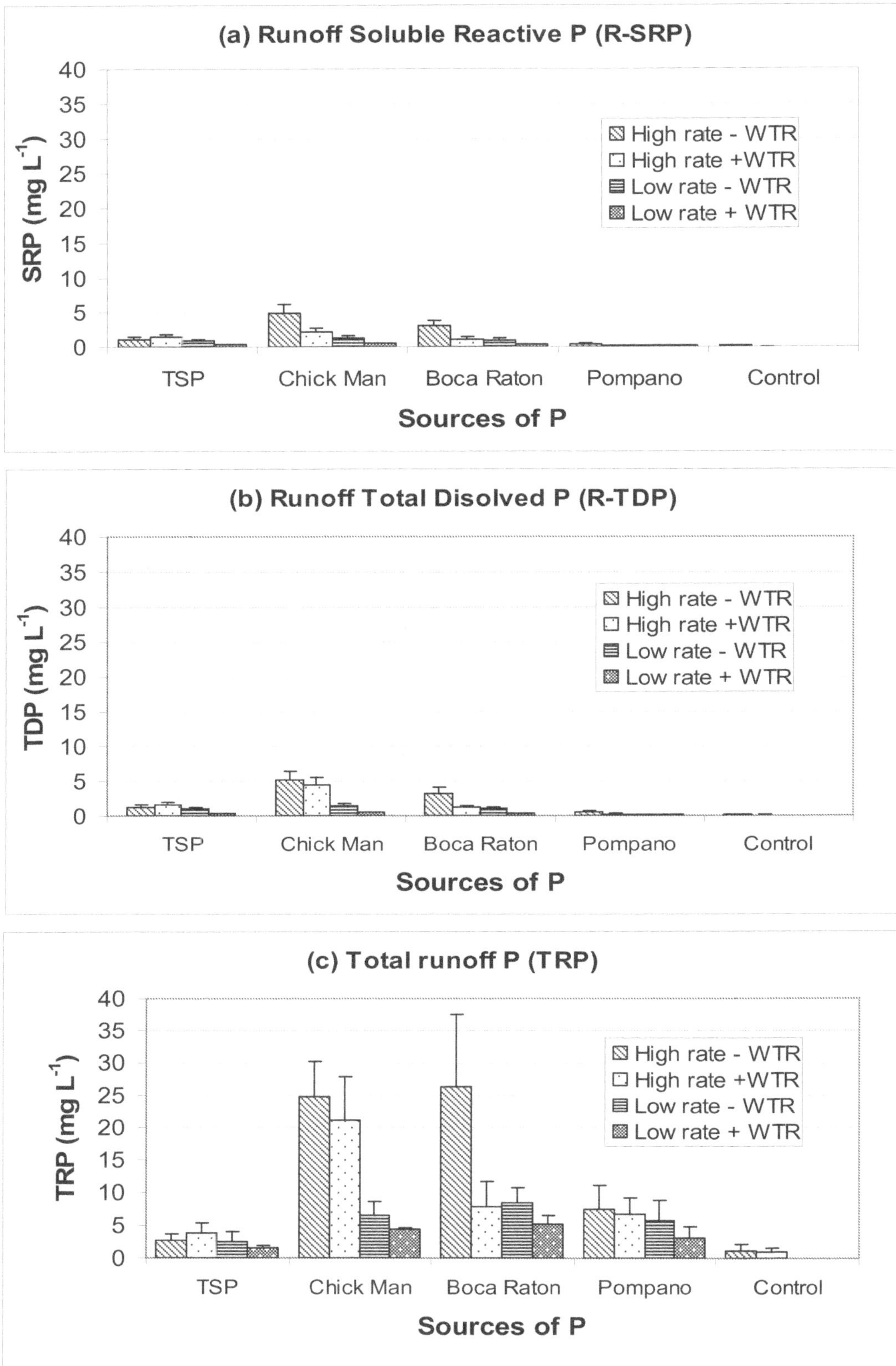

Figure 3-12. Forms of Runoff P for Various Treatments. Flow-weighted over three rainfall events; error bars represent standard deviations of three replicates.

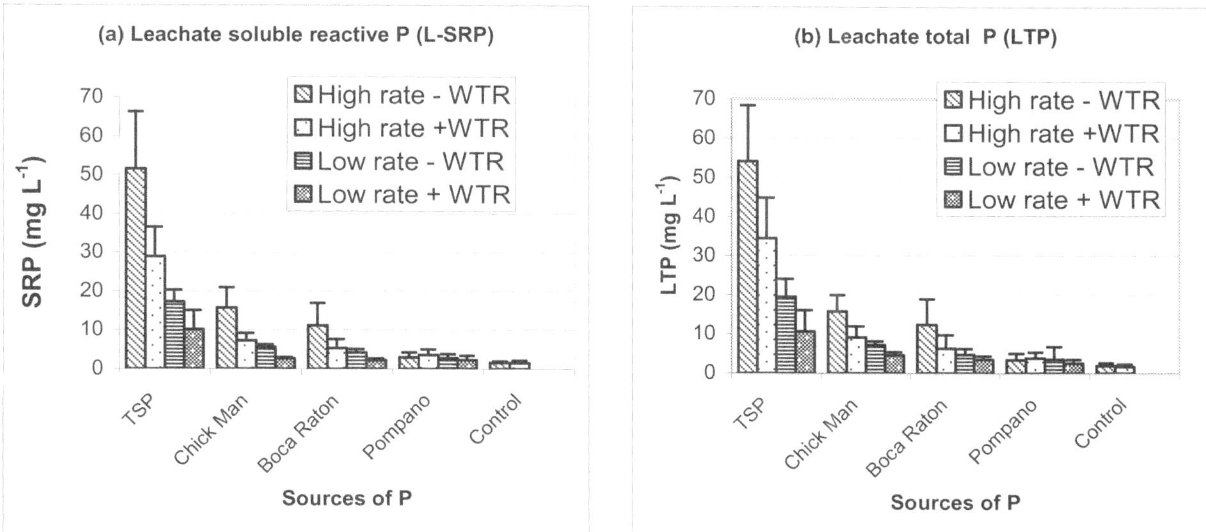

Figure 3-13. Forms of Leachate P for the Various Treatments. Flow-weighted over three rainfall events; error bars represent standard deviations of three replicates.

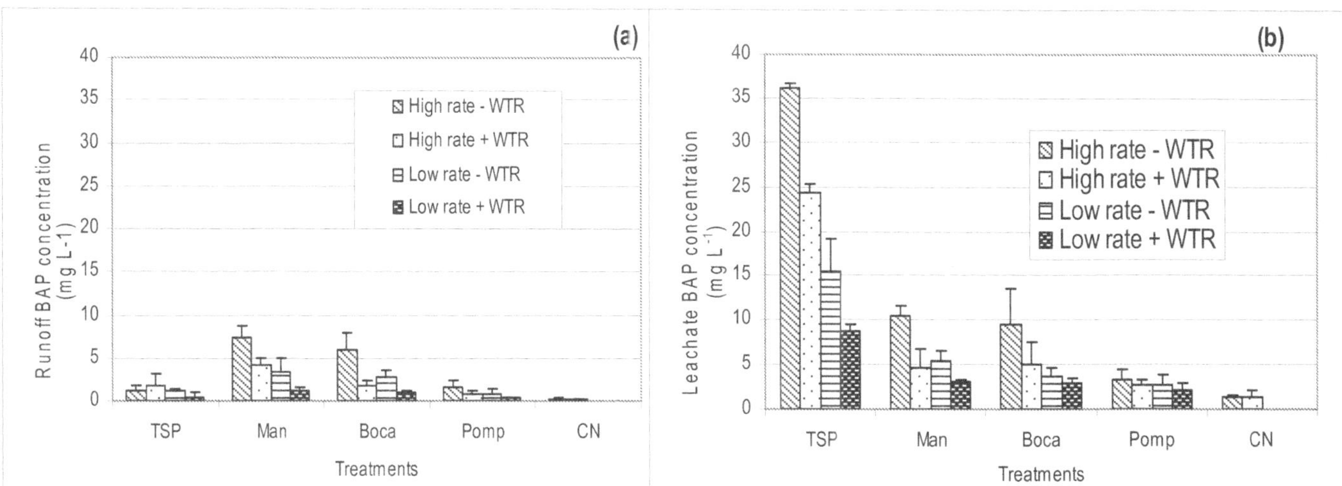

Figure 3-14. Flow-weighted Runoff (a) and Leachate (b) BAP Concentrations for the Various Treatments. Flow-weighted over three rainfall events; error bars represent standard deviations of three replicates.

To compare total P losses (runoff and leachate), flow-weighted concentrations were converted to masses by accounting for runoff and leachate volumes. Masses of BAP in runoff and leachate, and the sums, are shown in Figures 3-15 a, b, and c, respectively. Masses of all P forms in runoff and leachates are given in Appendix Table A-7.

Total BAP mass loss for fertilizer-P (TSP) was dominated by BAP mass loss in leachate, whereas BAP mass losses from the organic sources of P were roughly equal in runoff and leachate. Total BAP mass losses for the high rate of fertilizer-P (with or without WTR) were clearly greater than total BAP mass losses for the other treatments. Identification of differences between the BAP losses for the other treatments is complicated by significant P-source × source rate interactions (ANOVA, Appendix Table A-8). There was no source × WTR or rate × WTR interaction, however, so the significant effect of WTR addition was uncomplicated. Across sources and rates, addition of 1% (10 T A^{-1}, 22.4 Mg ha^{-1}) Al-WTR significantly reduced total BAP mass loss. Effects of WTR addition on total BAP loss for individual sources, source rates, or combinations of sources and rates were often not statistically significant, and likely reflected the minimal mixing and reaction time of WTR with P-sources allowed in the rainfall protocol. Thorough mixing of WTR with P-sources and extended reaction times promote more complete immobilization of soluble P and greater effects of WTR treatment on P leaching losses (Elliott et al., 2002; Silveira et al., 2006).

Total BAP mass losses for the various P-sources are presented in Figures 3-16 a and b for the high (N-based) and low (P-based) application rates, respectively. At both application rates, total BAP loss was less from the organic sources of P than from fertilizer-P. Differences in total BAP loss among the various organic sources were not statistically significant at either rate (Figure 3-16), but trends at the N-based rate (Figure 3-16a) suggest greater losses for sources with greater water-extractable P values. Thus, total BAP losses for poultry manure (water-extractable P = 4.6 g kg^{-1}) and Boca Raton biosolids (water-extractable P = 5.5. g kg^{-1}) are nearly the same, and losses from both sources tend to exceed losses with Pompano biosolids (water-extractable P = 1.2 g kg^{-1}). Total BAP losses from organic sources applied at P-based rates (Figure 3-16b) remained significantly less than losses from TSP, but differences among organic sources were less obvious.

Total BAP losses from organic sources of P applied at N-based rate (Figure 3-16a) were similar in magnitude (absolute values) to BAP losses from fertilizer applied at a P-based rate. Collectively, the BAP data support previous observations that biosolids-P applied at N-based rates is no more environmentally hazardous than applying fertilizer-P at P-based rates. There is less difference in environmental hazard (as estimated by total BAP mass) among the organic sources, but there is a tendency for sources with low water-extractable P values to lose less BAP than high water-extractable P materials. Addition of WTR reduces total BAP loss across P sources and rates.

3.3.4 Field Rainfall Simulation

To complement indoor rainfall runoff studies using packed soil boxes, a field rainfall simulator study was conducted on the Penn State research farm. Six treatments, consisting of two manures (dairy, poultry) and four biosolids (Philadelphia cake, Baltimore pellets, University cake, and Largo cake) were all loaded at 100 kg P ha^{-1} rate. The discussion here focuses on runoff P losses; detailed information on site and climatic conditions during the study can be found in Logan (2004).

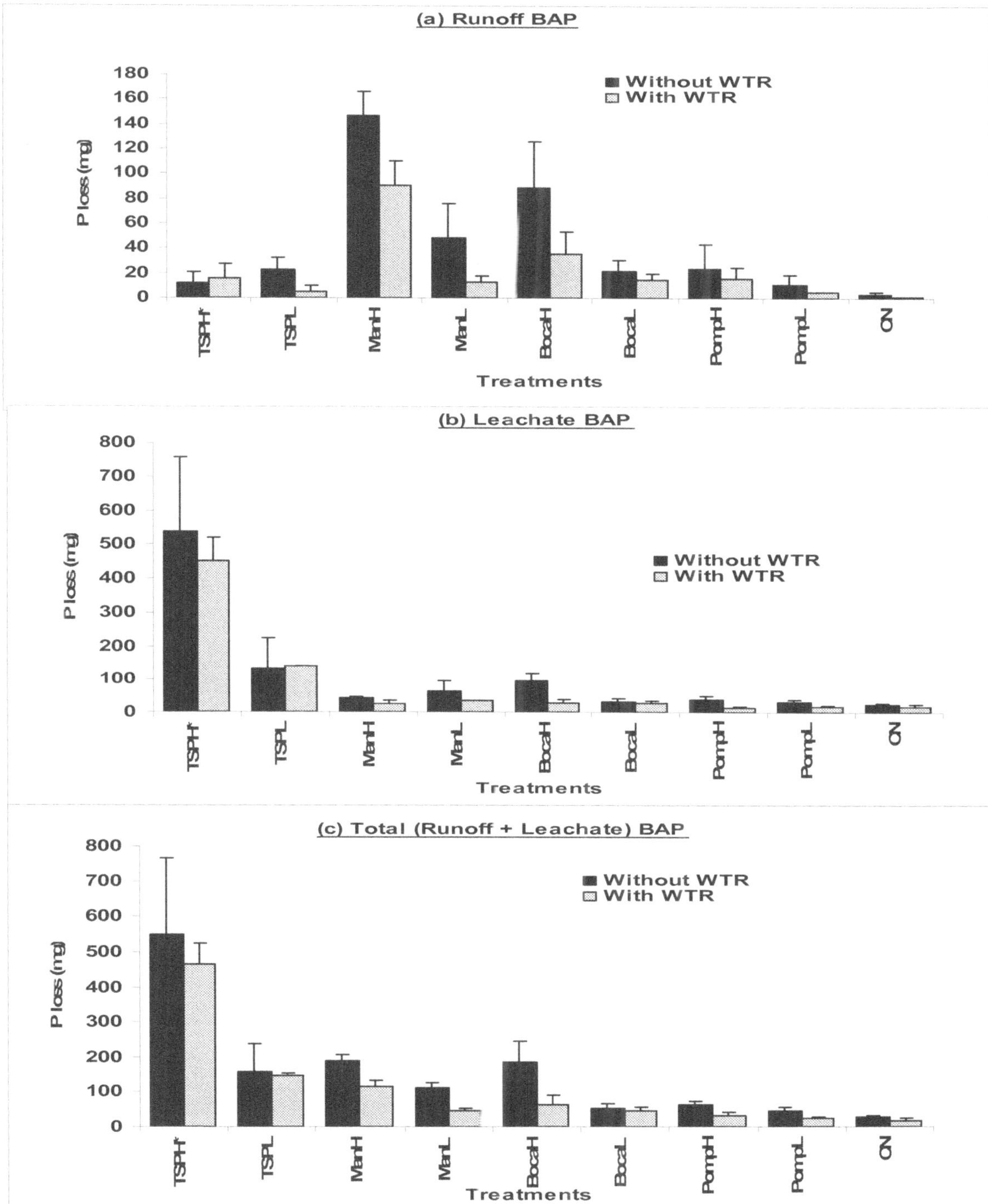

Figure 3-15. Cumulative Mass of Bioavailable P Lost During the Three Rainfall Events in the Leachates and the Runoff. Error bars represent standard error of three replicates.

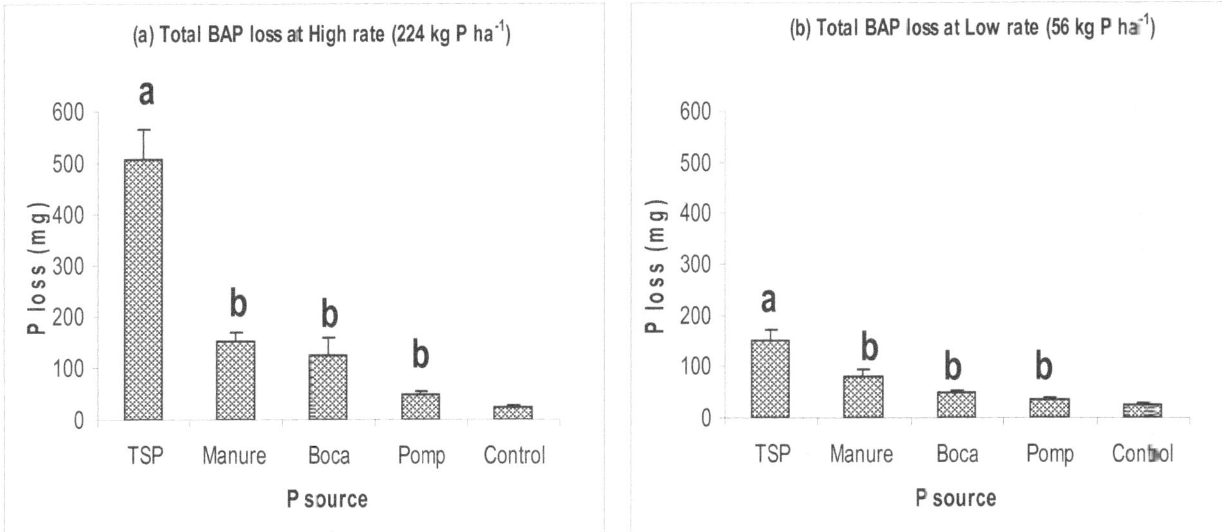

Figure 3-16. Total BAP Mass Lost During the Three Rainfall Events as Affected by the P Sources at (a) N-based and (b) P-based Application Rates. Treatments followed by the same letters are not different by the Scheffe multiple comparison at $P < 0.05$.

3.3.4.1 Soil and P Sources

The initial STP across plots (bulked) was 60 mg kg^{-1} (Mehlich-3 P). The STP concentration is slightly above optimum (30–50 mg kg^{-1}) for most commercial crops in Pennsylvania (Beegle, 2002). Even so, addition of fresh P sources was expected to mask effects of elevated STP levels on dissolved P in runoff (Andraski and Bundy, 2003).

Results of Mehlich-3 P and pH analyses for soil sampled inside the plots after the final rainfall simulation are shown in Table 3-5. The mean STP of the six sites after the final simulation was 134 mg kg^{-1}, with no significant difference between individual STP values at the six sites ($P = 1.000$). Assuming a bulk density of 1.33 g cm^{-3} and a depth of incorporation of 2.5 cm, an application of 100 kg P ha^{-1} will add 300 mg total P kg^{-1} to the soil. According to *The Illinois Agronomy Handbook*, a 1 unit of mass increase in STP (Bray) results when 4 units of mass of P are applied to the soil (Czapar and Simmons, 2002). The findings in this study seem to be consistent with this general behavior, as the STP after P source application (134 mg kg^{-1}) consists of approximately one-quarter of the applied P (75 mg kg^{-1}) in addition to the initial STP (60 mg kg^{-1}).

Table 3-5. STP and pH of Individual Plots after Final Rainfall-Runoff Simulation, 23 Days After P Source Application.

		Site 1 (UC)	Site 2 (PC)	Site 3 (BP)	Site 4 (LC)	Site 5 (PM)	Site 6 (DM)
Mehlich-3 P	Plot A	113	130	101	116	277	130
(mg kg^{-1})	Plot B	134	118	111	122	112	139
pH	Mean	5.6	5.6	5.7	5.7	6.3	6.5

UC, University cake; PC, Philadelphia cake; BP, Baltimore pellets; LC, Largo cake; PM, poultry manure; DM, dairy manure.

Soil pH did not seem to change with the application of biosolids; however, the pH increased in the manure-amended plots (Table 3-5). The mean soil pH (1:1 soil:water pH) among plots before P source application was 5.8. After the simulations, pH increased to 6.3 and 6.5 in the poultry manure and dairy manure plots, respectively.

Properties of the preapplied P source materials are shown in Table 3-6. The biosolids and manures had total solids content ranging from 14–34% (for the fresh manures and dewatered cakes) to 89% for the heat-dried and pelletized biosolids. Total Fe and Al concentrations in biosolids were significantly higher than in manures. Mean total Fe and Al concentrations for the biosolids were 51.1 and 22.0 g kg^{-1}, respectively, whereas the manures contained only 1.21 and 1.26 g kg^{-1}. The biosolids (except Baltimore pellets) and dairy manure are low in Ca (<5.9 g kg^{-1}) whereas the poultry manure is high in Ca, presumably due to Ca-enriched diets for eggshell production (Brandt et al., 2004). The relatively high Ca content of the Baltimore pellet material likely reflects the addition of lime during wastewater treatment.

Table 3-6. Characteristics of P Source Materials for the Field Simulation.

P Source	Label	Total Solids (%)	pH	Total Concentration (g kg^{-1})				Water-Extractable P (g kg^{-1})
				P	Fe	Al	Ca	
University cake	UC	14.11	7.19	33.17	49.89	11.84	5.45	0.09
Philadelphia cake	PC	25.44	7.74	27.10	91.39	30.35	5.87	0.16
Baltimore pellets	BP	89.25	6.92	29.09	53.06	30.64	19.33	0.25
Largo cake	LC	16.88	6.41	41.88	9.86	15.28	5.53	7.76
Poultry manure	PM	33.75	7.74	17.07	0.84	0.74	38.69	1.24
Dairy manure	DM	21.28	6.86	3.88	1.57	1.79	2.68	1.99

Water-extractable P was appreciably smaller for the conventional biosolids (0.09 to 0.25 g kg^{-1}), than for the BPR biosolids (7.8 g kg^{-1}). Dairy manure had a water-extractable P of ~2 g kg^{-1}, but poultry manure water-extractable P was only 1.24 g kg^{-1} (Table 3-6). In contrast, Kleinman and Sharpley (2003) reported water-extractable P values of 2 and 7 g kg^{-1} for dairy and poultry manure, respectively. In both studies, the poultry manure samples were collected from laying operations. In a survey of various livestock manures, Kleinman et al. (2005) reported mean values of 2.3 and 5.5 g kg^{-1} for water-extractable P and total P of dairy manure (bedded pack) and 4.9 and 25.6 g kg^{-1} for water-extractable P and total P of poultry manure (layer). Brandt et al. (2004) reported four layer poultry manures to have water-extractable P values of 2.7–5.8 g kg^{-1}. They also reported mean total P concentrations of dairy manure as 7.1 g kg^{-1}, much less than poultry manure (mean total P = 21.3 g kg^{-1}). Differences in handling, storage, feed, and litter content can cause widely varying total P content of individual manures (Brandt et al., 2004), which may explain the low total P content (3.88 g kg^{-1}) of the dairy manure. The data emphasize the highly variable nature of some P sources and the difficulty of obtaining representative samples of P sources containing "inert" materials (e.g., eggshells, feathers, wood chips, etc.).

3.3.4.2 Total Dissolved P in Runoff

Before application of P sources, the mean background total dissolved P concentration in runoff from the six sites was 0.29 mg L^{-1} (Figure 3-17). Background total P and particulate P concentrations were 0.56 and 0.27 mg L^{-1}, respectively. Phosphorus loss to surface runoff from the unamended field plots was nearly equally divided between particulate and dissolved P.

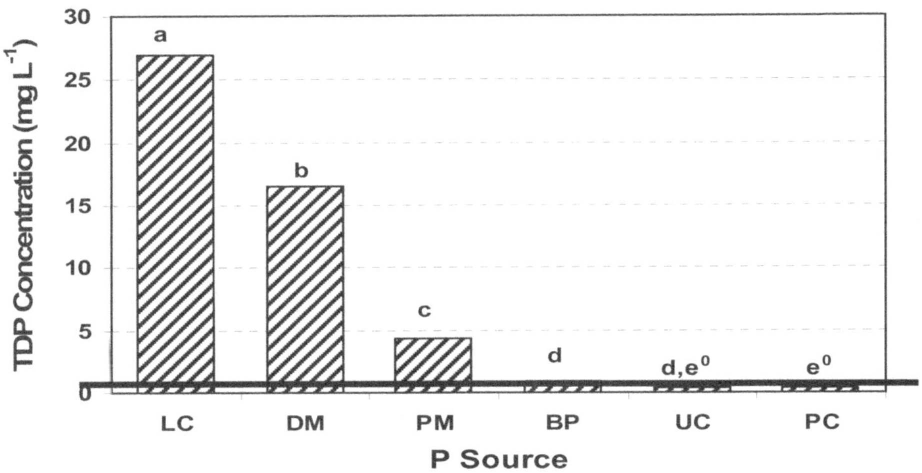

Figure 3-17. Mean Total Dissolved P in Runoff for the Different P Source Treatments for the Day 1 Field Rainfall Event. The horizontal line represents mean background total dissolved P before P-source application: Day 0 = 0.29 mg L^{-1}. Treatments followed by the same letter designations are not statistically different (α = 0.05). Superscript "0" indicates total dissolved P values not statistically different than background (pretreatment) value for the field.

A material-dependent trend was found when comparing total dissolved P loss in runoff to P source treatment for each rainfall event. Results and statistical analyses of total dissolved P for the three rainfall events are presented in Figures 3-17, 3-18, and 3-19, for days 1, 2, and 22, respectively. Dissolved P concentrations in runoff tended to decrease with subsequent rainfall events (with the exception of those P sources not showing significant P losses). This is consistent with the finding that the majority of total dissolved P loss from biosolids and manures occurs during the first runoff event (Kleinman and Sharpley, 2003; Pierson et al., 2001; Sims et al., 2003; Withers and Bailey, 2003).

Differences in total dissolved P values between the P sources were consistent for the first two rainfall events (Figures 3-17 and 3-18): Largo cake > dairy manure > poultry manure > Baltimore pellets ≈ University cake ≈ Philadelphia cake. This follows the same trend as P source water-extractable P values: Largo cake (7.76 g kg^{-1}) > dairy manure (2.00) > poultry manure (1.24) > Baltimore pellets (0.25) ≈ University cake (0.09) ≈ Philadelphia cake (0.16). Water extractable P was correlated with total dissolved P in runoff (r = 0.94 and 0.92, for Days 1 and 2, respectively). For Day 1, Largo cake had a very high total dissolved P concentration (27 mg L^{-1}); dairy manure also had a relatively high total dissolved P value (16.5 mg L^{-1}) and poultry manure followed at 4.4 mg L^{-1}.

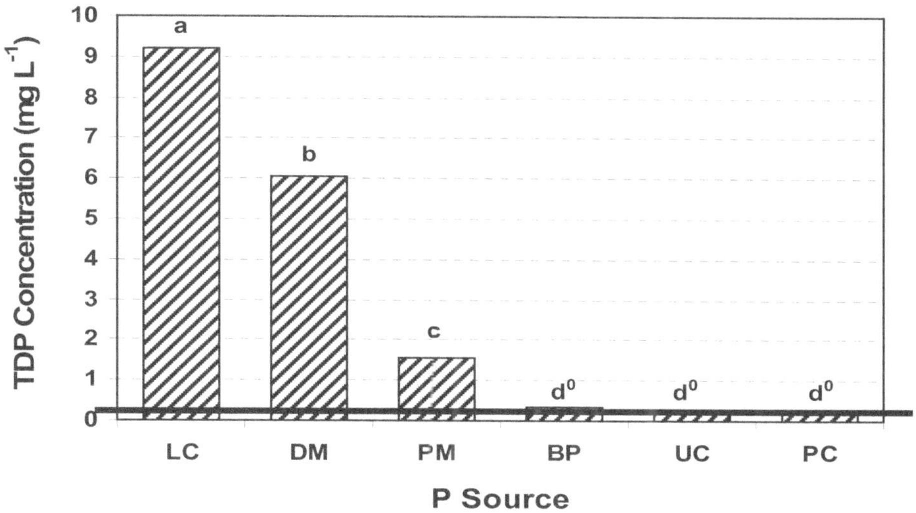

Figure 3-18. Mean Total Dissolved P in Runoff for the Different P Source Treatments for the Day 2 Field Rainfall Event. The horizontal line represents mean background total dissolved P before P-source application: Day 0 = 0.29 mg L^{-1}. Treatments followed by the same letter designations are not statistically different ($\alpha = 0.05$). Superscript "0" indicates total dissolved P values not statistically different than background (pretreatment) value for the field.

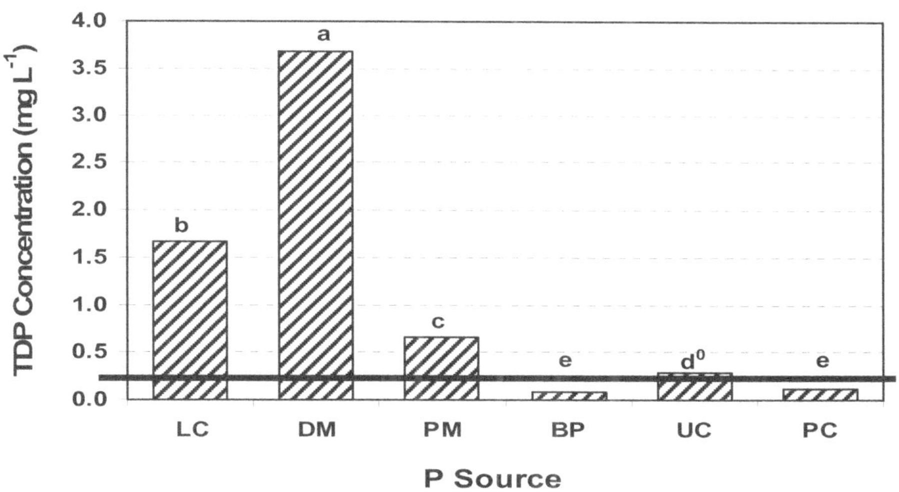

Figure 3-19. Mean Total Dissolved P in Runoff for the Different P Source Treatments for the Day 22 Field Rainfall Event. The horizontal line represents mean background total dissolved P before P-source application: Day 0 = 0.29 mg L^{-1}. Treatments followed by the same letter designations are not statistically different ($\alpha = 0.05$). Superscript "0" indicates total dissolved P values not statistically different than background (pretreatment) value for the field.

Kleinman and Sharpley (2003) documented higher P concentrations in runoff from soil amended with poultry manure than with dairy manure due to greater water-extractable P content of poultry than dairy manure at equivalent P loads. Though plots amended with dairy manure in this experiment lost significantly more total dissolved P in runoff than plots amended with poultry manure, water-extractable P was also higher in the dairy manure material than in the poultry manure, which again supports the hypothesis that water-extractable P can accurately predict total dissolved P in runoff.

Research has shown that biosolids-P tends to be less susceptible to removal in runoff than manure P. Withers et al. (2001) observed a lower risk of P transfer to surface runoff from agricultural land amended with liquid and dewatered biosolids than from land amended with liquid manure or inorganic P fertilizer at similar P loadings.

Several biosolids produced runoff total dissolved P concentrations comparable to the preapplication (Day 0) total dissolved P (0.29 mg L^{-1}) during the first two rainfall events (Figures 3-17 and 3-18). Statistically, runoff total dissolved P from the University cake and Philadelphia cake plots was no different than the total dissolved P of the unamended plots on Day 1 (P = 0.922 and 1.00, respectively); and University cake, Philadelphia cake, and Baltimore pellets runoff total dissolved P was no different than that of the unamended plots on Day 2 (P = 0.990, 0.999, and 0.996, respectively). Essentially, adding these conventional biosolids to the field plots had no effect on the concentration of total dissolved P in runoff. This is likely due to the high Fe and Al content of the biosolids, which results in formation of Al- and Fe-bound P compounds of low water solubility. Such compounds represent a relatively inert P pool that is not likely to contribute substantial amounts of dissolved P to the runoff.

A slightly different trend in total dissolved P values among P sources was observed during the Day 22 rainfall event: dairy manure > Largo cake > poultry manure > University cake > Baltimore pellets = Philadelphia cake (Figure 3-19). Largo cake (BPR biosolids) yielded a total dissolved P concentration significantly lower than that of dairy manure. The Largo cake resulted in very high total dissolved P concentrations for the initial rainfall event, but this decreased more rapidly than the total dissolved P levels in the manure treatments. It is possible that the extremely soluble nature of the Largo cake material allowed total dissolved P to easily infiltrate into the soil. Pote et al. (2001) observed that dissolved P concentration in runoff was inversely correlated with the tendency of the soluble portion of the applied P source to infiltrate into the soil. On the other hand, there was an 8.4% loss of applied P for Largo cake compared to 4.9% for dairy manure on Day 1, and 2.8% for Largo cake and 1.7% for dairy manure on Day 2. Thus, there was a more rapid reduction in the pool of total dissolved P available for transport with the Largo cake material.

The conventional biosolids continued to yield runoff total dissolved P concentrations comparable to the background total dissolved P on Day 22. The University cake runoff total dissolved P was the same as the preapplication total dissolved P (P = 1.00) and the Baltimore pellets and Philadelphia cake runoff total dissolved P values were smaller than the preapplication total dissolved P. Although Maguire et al. (2000b) suggested that biosolids high in reactive Fe and Al could reduce P runoff by increasing the P sorption capacity of soils amended with biosolids, the present results do not necessarily indicate that these biosolids sorbed P from the soil. As no control plot was tested concurrent to the research plots, it is possible that total dissolved P from an unamended plot would have been even lower than the preapplication value if subjected to the three intense rainfall events used in this study and the natural precipitation impinging on the plots between Days 2 and 22.

3.3.4.3 Water-Extractable P and P Transport

The P loss from agricultural soils amended with organic sources of P is related to the water extractable P concentration of the source material (Brandt and Elliott, 2003; Brandt et al., 2004; Kleinman and Sharpley, 2003; Kleinman et al., 2002b; Sharpley and Moyer, 2000). Data generated here support the relationship (Figure 3-20). Values of total dissolved P increased proportionally with P source water-extractable P for the first two rainfall events. Water-extractable P concentrations were lowest for conventional biosolids Baltimore pellets, University cake, and Philadelphia cake (< 1 g P kg^{-1}), paralleling the negligible total dissolved P concentrations in runoff for each rainfall event. Poultry and dairy manures had greater water-extractable P values (1.2 and 2.0 g P kg^{-1}, respectively), and Largo cake had the greatest water-extractable P (7.8 g P kg^{-1}).

For the first rainfall event, the total dissolved P vs. water-extractable P coefficient of determination (r^2) was 0.88. This indicates that 88% of the change in total dissolved P of those P sources is explained by variations in water-extractable P values. Day 2 also had a high correlation ($r^2 = 0.85$) demonstrating that water-extractable P continued to be a good predictor of total dissolved P in runoff. The Day 22 data are also linearly correlated, but the low r^2 value (0.18) reflects the nearly horizontal trendline for the data. Regardless of initial P source water-extractable P concentration, runoff total dissolved P was generally less than 4 mg L^{-1} after 22 days, reflecting extraction of soluble P from all sources.

The relationship between P source water-extractable P and total dissolved P in this field simulation experiment is similar to that observed in an indoor rainfall simulation study with some of the same materials (Section 3.3.2.3). Although conditions varied between these two experiments (i.e., bare soils vs. grassed plots, varying total P application rates), the relationships between water-extractable P and total dissolved P were similar in both experiments. The first rain event of both field and indoor rainfall simulations yielded similar slopes (3.49, 3.16) and coefficients of determination (0.88, 0.83) between total dissolved P and water-extractable P. This

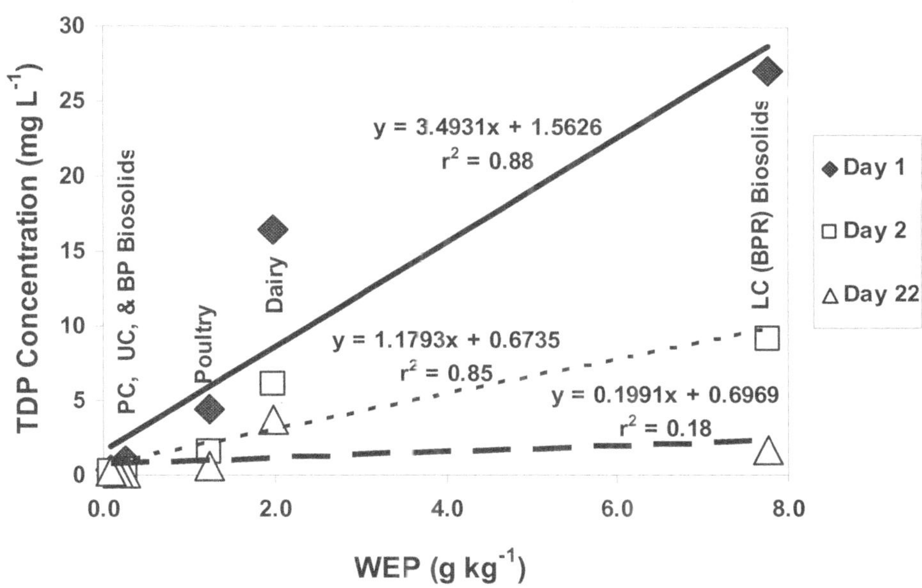

Figure 3-20. Mean Total Dissolved P Concentrations in Runoff as a Function of P Source Water-Extractable P (WEP) for Each Field Runoff Event.

suggests that indoor simulations are reasonably representative of P movement in surface runoff under field conditions. Kleinman et al. (2004) suggested that both field plots and packed indoor boxes can be used to produce comparable coefficients for P site assessment indices despite differences in rainfall, hydrology, and erosion.

The dramatic differences in water-extractable P values between biosolids and manures can most likely be attributed to different WWTP processes and metal treatments. Most treatment facilities that transport biosolids considerable distances use mechanical dewatering (belt presses, centrifugation, vacuum filtration) to remove liquid from the biosolids, subsequently removing much water-soluble P entrained in the filtrate (Brandt et al., 2004).

Iron and aluminum additions to biosolids are equally important factors explaining the difference between biosolids and manures. The National Pollution Discharge Elimination System permit limits effluent P to control eutrophication in receiving waters for many municipalities. The most common method for removing dissolved P from wastewater is precipitation of metal phosphates through addition of Fe and Al salts. These salts are also used for odor control and conditioning agents in dewatering. As WWTPs differ in operational processes, biosolids will have variable levels of Fe and Al from different plants, depending on treatment objectives

Given the usefulness of source- water-extractable P as an indicator of P loss in runoff (Kleinman et al., 2002b), runoff total dissolved P should be correlated with the Fe and Al contents of the P sources. The relationship between total dissolved P and P source molar Fe plus Al concentration is shown in Figure 3-21. The relationship has been modeled as a power function (r^2 = 0.39).

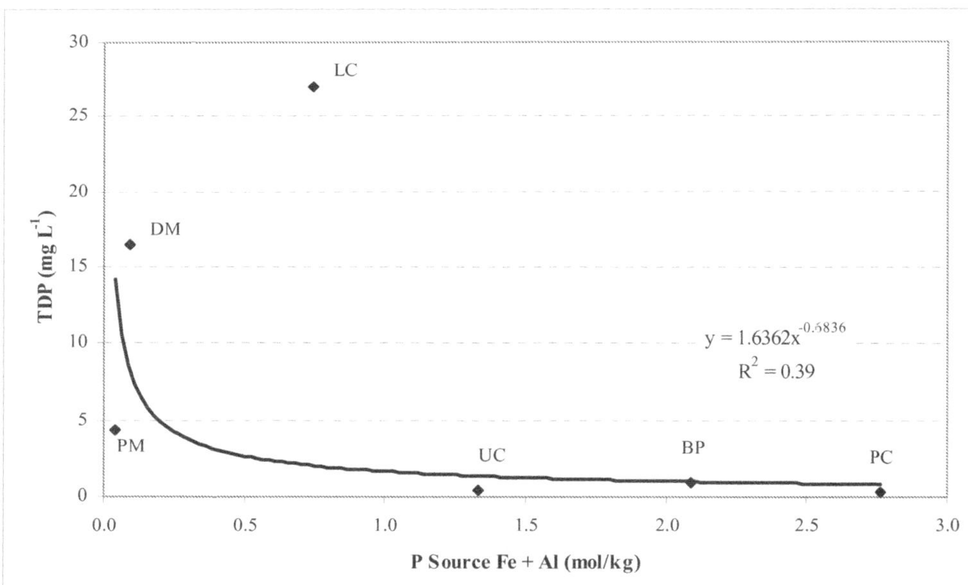

Figure 3-21. Day 1 Total Dissolved P in Field Simulation Runoff for Each P Source as a Function of Total Molar Fe plus Al Concentrations of the P Sources. DM, dairy manure; PM, poultry manure; LC, Largo cake; BP, Baltimore pellet; UC, University cake; PC, Philadelphia cake.

Runoff collected from conventional biosolids treated with Fe and Al had minimal total dissolved P concentrations in runoff, most of which were comparable to pre-application total dissolved P levels (0.29 mg L^{-1}). The Baltimore pellet material had a high total P concentration (29.1 g kg^{-1}), but contained a substantial amount of Fe (53 g kg^{-1}) and resulted in relatively low runoff total dissolved P. For each runoff event, Baltimore pellets had total dissolved P concentrations similar to background total dissolved P levels (Day 1 slightly higher, Day 2 no different, and Day 22 slightly less than Day 0). Philadelphia cake and University cake also contained high Fe concentrations (72.3 and 9.1 g kg^{-1}, respectively) and had runoff total dissolved P concentrations no greater than the background total dissolved P for each rain event. The high metal concentrations likely increased the P-sorbing potential of the plots and mitigated release of P into runoff, as suggested by Maguire et al. (2000). However, Fe plus Al is not the sole determinant of P loss to runoff, as discussed earlier. Largo cake does not seem to fit well in this relationship between Fe plus Al concentration and runoff total dissolved P. This is most likely because of the BPR wastewater processing and the limited addition of metal salts which collectively cause the material to act differently than conventionally treated biosolids. Processes for BPR ultimately lessen the amount of sludge produced when compared to conventional P precipitation which adds extra mass (Fe and Al salts) to sludge. Biosolids from BPR plants are typically low in Fe and Al, have higher soluble P (Brandt et al., 2004; Kyle and McClintock, 1995), and greater P-leaching potential (Elliott et al., 2002).

3.3.4.4 Total P in Runoff

Results and associated statistical analyses of runoff total P (particulate P + total dissolved P) for each P source during the three field rainfall events are presented in Figure 3-22. Most of the total P lost in runoff from the manures and biosolids was in particulate form, with exception of the Largo cake and Baltimore pellet materials. Though the total dissolved P of Largo cake was nearly double that of the manures for the first two rain events, total P concentrations of Largo cake, dairy manure, and poultry manure were statistically the same for Days 1 and 2 ($P > 0.27$ and $P > 0.083$, respectively). By Day 22, total P loss from dairy manure was statistically greater than that of the Largo cake and poultry manure treatments ($P = 0.001$). Infiltration of soluble P may have been instrumental in allowing a rapid decrease in total P (93% dissolved P) concentration of the Largo cake material, or the dissolved P reservoir in the Largo cake material was being depleted.

Field simulation results suggest that the nature of the P source (cake, slurry, pellets) influences the erosion potential and, consequently, the total P lost during a rainfall event. McDowell and Sharpley (2003) similarly found that more total dissolved P was transported from dairy manure than poultry manure-amended soil, but more total P was lost from the poultry manure soil than the dairy manure soil due to a greater contribution of particulate P. Poultry manure may be more susceptible to erosion than dairy manure or it may contain a substantially greater amount of particulate P.

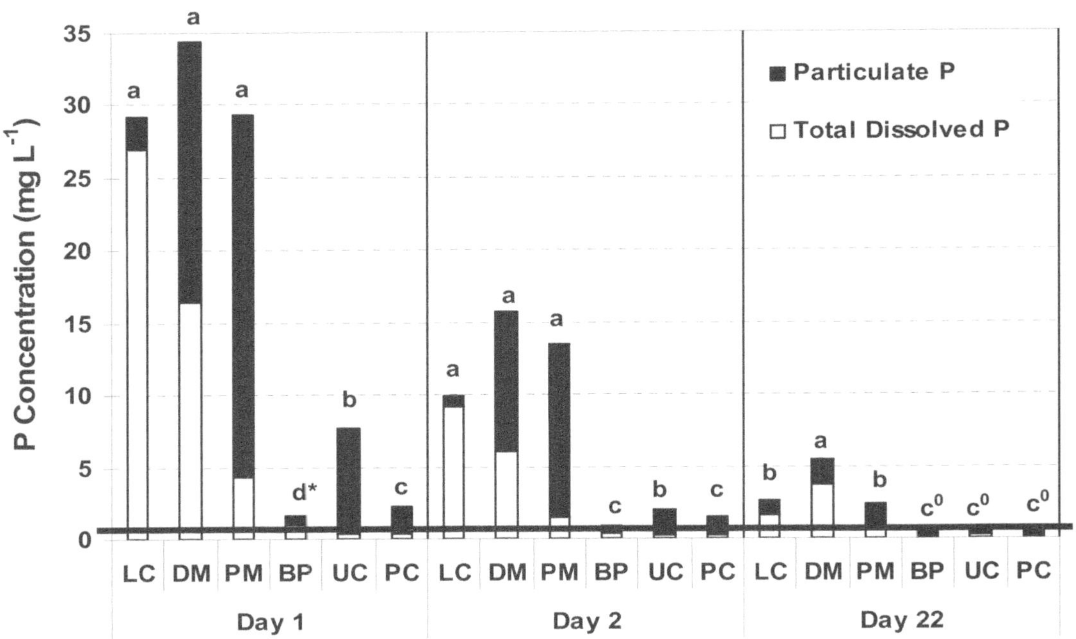

Figure 3-22. Mean Total P (Total P = Particulate P + Total Dissolved P) in Runoff for the Different P Source Treatments for Each Field Rainfall Event. Treatments followed by the same letter designations within Day (1, 2, 22) are not statistically different (α= 0.05). Superscript "0" indicates total P values not statistically different than the background (pretreatment) total P concentration for the field represented by the horizontal line at 0.56 mg L^{-1}. LC, Largo cake; DM, dairy manure; PM, poultry manure; BP, Baltimore pellets; UC, University cake; PC, Philadelphia cake.

The anaerobically digested cake biosolids (University cake, Philadelphia cake) yielded much more particulate P than total dissolved P in runoff (Figure 3-22). Elevated particulate P values increased the total P concentrations above those of the pre-application Day 0 total P concentration (0.56 mg L^{-1}) for the first two rainfall events. By Day 22, however, total P concentrations in University cake and Philadelphia cake runoff were not distinguishable from the background total P. Because of the low solids (12 %) of University cake, the material was easily scoured from the surface causing more suspension of material in runoff.

Total P concentrations in runoff from the heat-dried and pelletized Baltimore pellets biosolids treatment were significantly greater than the background total P only on Day 1 (P = 0.025). Total P in runoff from the Baltimore pellets consisted mostly of total dissolved P on Day 1, whereas on Days 2 and 22, runoff total P was predominantly particulate P. During weekly collection of soil samples, Baltimore pellet materials were observed to have been washed from the vegetation and were sitting on the soil surface. Erosion losses of the Baltimore pellets were negligible as the rainfall did not easily dislodge the pellets trapped below the vegetation.

3.3.4.5 Field Simulation Conclusions

For the P sources that showed substantial P losses, total dissolved P concentrations decreased with each rainfall event, consistent with previous findings that most P is loss in the initial rainfall event. For the first two rainfall events, a similar trend in total dissolved P among the P sources was observed: Largo cake > dairy manure > poultry manure > Baltimore pellets ≈

University cake ≈ Philadelphia cake. This also follows the order of P source water-extractable P (in g kg^{-1}): Largo cake (7.76) > dairy manure (2.00) > poultry manure (1.24) > Baltimore pellets (0.25) ≈ University cake (0.09) ≈ Philadelphia cake (0.16). The biosolids generated via a BPR process (Largo cake) had exceptionally high water-extractable P values and P runoff potential. Three of the biosolids (Baltimore pellets, University cake, Philadelphia cake) exhibited total dissolved P levels (Days 1 and 2) that were indistinguishable from the unamended plots.

For the Day 22 rainfall event, the Largo cake material resulted in total dissolved P levels significantly lower than those for dairy manure. The BPR biosolids had exceptionally high total dissolved P levels initially (Day 1 and 2), but these levels seemed to diminish more rapidly than the manure treatments. Most of the total P (total dissolved P plus particulate P) lost from the Largo cake material was in a dissolved form, in contrast to the manures and other biosolids. For the Day 1 and 2 rainfall events, the total P losses for the Largo cake, dairy manure, and poultry manure were not statistically different. By Day 22, the total P loss from the dairy manure was statistically greater than the Largo cake and poultry manure. The data suggest that the nature of the P source (cake, pellets, slurry) influences the potential for scouring from the surface and, in turn, the total P lost during a rainfall event.

The total dissolved P levels for the Baltimore pellets and Philadelphia cake materials (Day 22) were statistically lower than the control plot levels. These two biosolids have very high Fe contents (Baltimore pellets = 53 g kg^{-1}; Philadelphia cake = 91.4 g kg^{-1}). Because there were no control plots which received the same amount of rainfall as treatments, it cannot be concluded unequivocally that biosolids addition reduced P losses. However, other researchers (Maguire et al., 2002b) have suggested that addition of these high-Fe biosolids increase the P-sorbing potential of soils, thus reducing the release of P into runoff.

3.3.5 Application Method Simulation

3.3.5.1 Soil and P Sources

Previous project rainfall simulations were conducted with surface applied P sources. From the potential impact on water quality perspective, this represents a worst-case scenario. If a P source is mixed in and equilibrated with the soil, sorption of P by soil surfaces is expected to lessen the amount of P susceptible to runoff. The Largo pelletized material, previously shown to have high runoff potential (Figure 3-11), was both surface applied on, and incorporated into, the Hagerstown soil (Table 3-3) in this experiment. The total P application rate was 100 kg P ha^{-1}. Besides the unamended soil-only control, the Philadelphia cake biosolids and dairy manure were surface applied and served as reference points for comparison to the other indoor runoff studies.

3.3.5.2 Runoff P Results

Figure 3-23 shows the runoff total dissolved P and total P values for the treatments ranked from highest (Largo pellets surface applied) to lowest (control) runoff total dissolved P. Highest runoff total P and total dissolved P values were exhibited by the surface applied Largo pellets and dairy manure. In a previous rainfall simulation, these two P sources also showed significant P loss at the 100 kg P ha^{-1} rate (Figure 3-4). Runoff P was not statistically different for the two materials applied at the same total P rate, in contrast to the runoff P losses when these P sources were applied at a constant plant-available N rate (Figure 3-7a). Both total dissolved P and total P values of the Philadelphia cake were statistically similar to the control values. In the

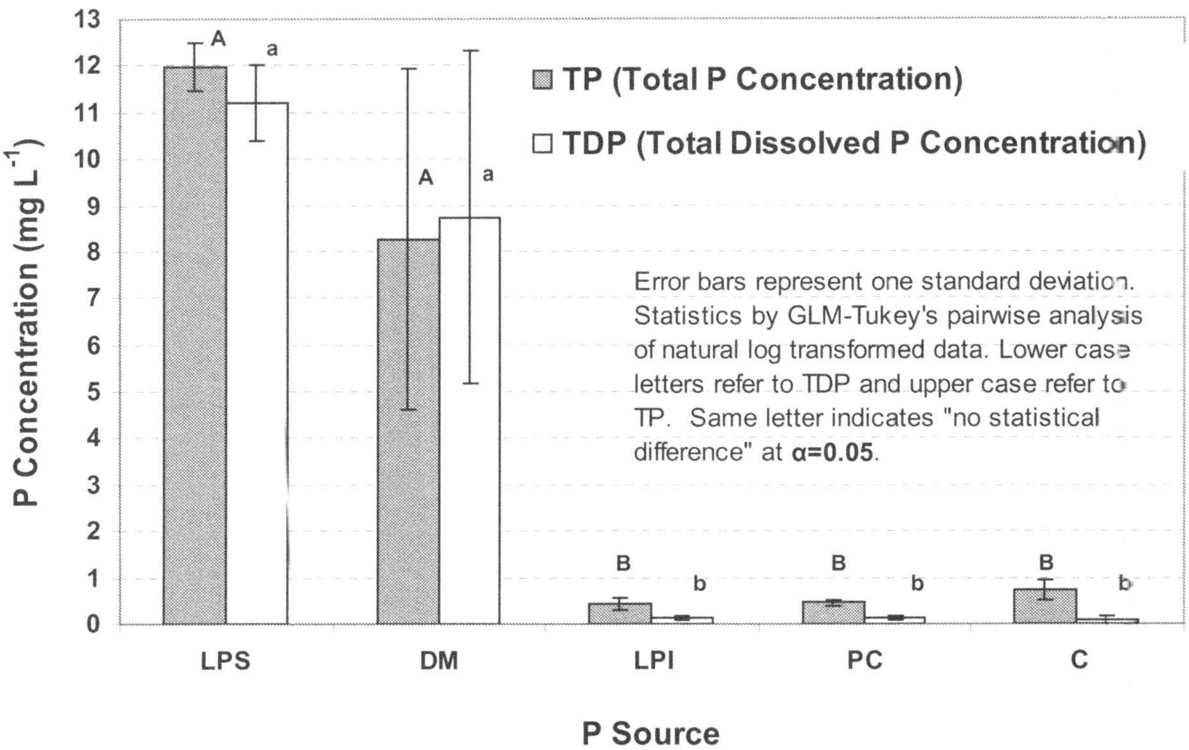

Figure 3-23. Runoff P (Total P and Total Dissolved P) from Largo Pellets Surface Applied (LPS) and Incorporated (LPI) Compared to Surface-Applied Dairy Manure (DM), Philadelphia Cake (PC), and the Unamended Soil Control (C).

variable P loading simulation (Figure 3-7a), runoff total dissolved P values were statistically similar for the Philadelphia cake and control treatments, but the total P loss for Philadelphia cake was greater than the control.

Incorporation dramatically reduced the runoff P losses from the Largo pellets (labeled LPI in Figure 3-23). Indeed, incorporation resulted in runoff P losses (both total dissolved P and total P) that were not statistically different that the soil-only control or the high-Fe Philadelphia cake biosolids. This finding is in contrast to that of Penn and Sims (2002), who found that total dissolved P and total P values from an incorporated biological N-removed biosolids were the highest among all the biosolids tested and were statistically higher than control levels. Penn and Sims (2002) used two soils, Elsinboro (silty clay loam) and Woodstown (sandy loam), with Mehlich 1-P levels of 21 and 124 mg kg^{-1}, respectively. The Hagerstown soil had a Mehlich 3-P of 95 mg kg^{-1} (Table 3-3), so it is unlikely that the different results of these two studies could be explained by the pre-application soil P content. The biological N-removed material used by Penn and Sims (2002) had a water-extractable P of 8.1 g kg^{-1} (10:1water-to-biosolids extraction ratio) compared to 4.55 g kg^{-1} (200:1 extraction ratio) for the Largo pellets. The total P loading rate used by Penn and Sims (2002) was 200 kg P ha^{-1}, twice the value used in the present study. The greater water-extractable P and total P application rate for the BPR material in the Penn and Sims (2002) study could explain why runoff P was affected despite incorporation of the biosolids into the soil.

Because Penn and Sims (2002) did not compare P losses between surface applied and incorporated biosolids, conclusions regarding the impact of incorporation on P runoff loss cannot be made from that study. However, the results from the present study clearly show the dramatic impact incorporation can have on runoff P concentration. This could have important practical significance for addressing land spreading of BPR biosolids, expected to exhibit exceptionally high P loss potential relative to other types of biosolids. For incorporation to be an effective P loss mitigation strategy, the soil must have appreciable P-sorbing capacity.

CHAPTER 4.0

SIGNIFICANCE OF FINDINGS AND FUTURE WORK

4.1 Introduction

Phosphorus continues to be at the forefront of biosolids-related issues that will affect the long-term sustainability of land-based recycling programs. Concern over water quality impacts from agricultural watersheds has elicited various national and state initiatives. The Unified Strategy for Animal Feeding Operations (USDA/U.S. EPA, 1999) requires each state Natural Resources Conservation Service (NRCS) to address P in nutrient management practice standards (Code 590). Most states have chosen a P indexing tool in updating Code 590 standards. The P index is preferred to an STP threshold approach because it addresses both source and transport of P and is predicated on the principle that significant P loss requires the coexistence of a labile P source *and* a viable transport pathway (Figure 4-1). Although efforts so far have targeted manures and commercial fertilizers, P-based management will eventually apply to *all* land-applied residuals including biosolids.

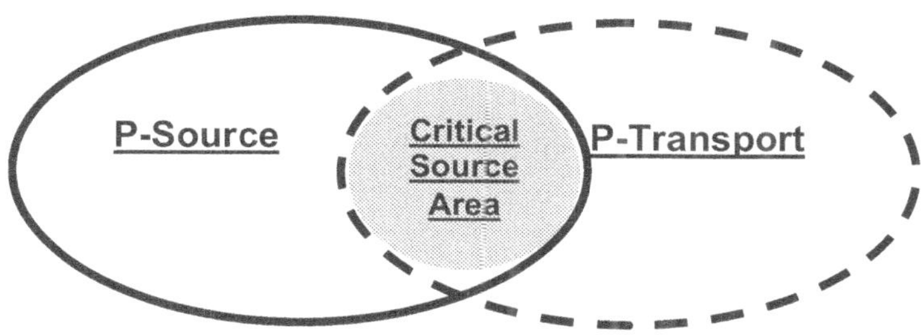

Figure 4-1. Venn Diagram Illustrating Basic P Index Concept—Critical P Source Areas Occur Only Where Elevated P-Source and Active P-Transport Conditions Coexist.

Most state P indices do not explicitly mention biosolids. This is unfortunate because the nature and transport potential of P depends critically on whether the P originated from manure, biosolids, or commercial fertilizers. Although there are some elements of commonality in state P indices, there are many different versions now in use. Coale et al. (2002) predicted that P indices will be deployed as regulation long before they are objectively validated. Because biosolids recycling is already highly regulated, there is even greater danger that imposition of regulatory restrictions will occur before the impact on land application programs is fully appreciated.

This chapter first briefly examines P management strategies using a P site index approach. Several necessary principles are then proposed for P management standards to promote land application as a sustainable biosolids management strategy. Reference is made to specific approaches used by individual states to address land application of biosolids based on P. The researchers highlight the substantial challenges that P-based nutrient management will impose on biosolids recycling programs for many municipalities. The chapter concludes with a discussion of critical gaps in knowledge and research needed to assure the long-term sustainability of land-based biosolids recycling under P nutrient management

4.2 Biosolids Recycling Under P-Based Management

4.2.1 The P Index Concept

The original P index (Lemunyon and Gilbert, 1993) was a field-scale tool designed to use readily available information to identify vulnerable sites where P-loss reduction efforts should be focused. The index contained five source and four transport factors (Table 4-1). Transport factors described soil and irrigation erosion, runoff, and distance from the watercourse. Source factors addressed soil-P status, P-source inputs (fertilizer and manures) for different crop and soil management practices. The factors were multiplied by weighting factors and the products summed to arrive at the final P site index value.

Individual states customized the original index to address local climate, hydrology, and geomorphology conditions. For example, because the original index had no capability to assess P leaching losses to shallow groundwater, the Maryland index was expanded to account for P leaching in its coastal plain soils (Coale et al., 2002). In the Pennsylvania index, irrigation erosion was eliminated, while subsurface drainage, modified connectivity, and manure P availability were added. Various weighting factors and arithmetic operations are used in P indices to reflect the interrelationships and relative importance of the parameters to P loss.

Although the details vary, most states have established interpretive categories that define nutrient management practices based on a site's P index value. For example, in Pennsylvania, four management categories are defined (low, medium, high, and very high). For the low and medium categories, soil application rates for organic by-products are determined using nitrogen (N)-based budgets. This is the procedure currently in place for biosolids, as CFR 40 Part 503 requires application at the "agronomic rate to provide the amount of *nitrogen* needed by the food crop, feed crop, fiber crop, cover crop, or vegetation grown on the land." The high category dictates careful management of P, with soil application rates of organic by-products restricted to just meeting the P requirement of the crop. This has important ramifications for biosolids management, because typical biosolids-P levels translate into application rates that are ~5–10 times lower than current practice. The very high category dictates that no additional P should be applied to the site.

4.2.2 Principles for Sustainable Biosolids Land Application

Sharpley et al. (2003) outlined the development and current status of P indices in the United States. They summarize P indices for 41 of the 47 states (as of 2003) using this approach to target P management. Initial efforts to develop P site indices focused on minimizing P loss from areas of intensive crop and livestock production. Besides inorganic P fertilizer, the other principal input of P was livestock manures. Yet the intention of the NRCS is that P management be applicable to *all* P sources put on the land. Biosolids have not been systematically addressed in most state P indices or assumed to be the same as manures. However, the solubility, bioavailability, and transport potential of biosolids-P is often quite different than that of typical livestock

Table 4-1. Original P Index, Version 1.0. Reprinted with permission of Lemunyon and Gilbert, 1993.

	Phosphorus loss rating (Value)				
Site characteristic (weighting factor)	**None (0)**	**Low (1)**	**Medium (2)**	**High (4)**	**Very high (8)**
Soil erosion (1.5)	Not applicable	<11.2 Mg/ha (<5 tons/A)	11.2–22.4 Mg/ha (5–10 tons/A)	22.4–33.6 Mg/ha (10–15 tons/A)	>33.6 Mg/ha (>15 tons/A)
Irrigation erosion (1.5)	Not applicable	Tailwater recovery or QS<6 for very erodible soils or QS<10 for others.	QS>10 for erosion resistant soils	QS>10 for erodible soils	QS>6 for very erodible soils
Runoff class (0.5)	Negligible	Very low or Low	Medium	High	Very high
Soil test P (1.0)	Not applicable	Low	Medium	High	Excessive
P fertilizer application rate (0.75)	None applied	1–34 kg P_2O_5/ha (1–30 P_2O_5 lb/A)	35–101 kg P_2O_5/ha (31–90 P_2O_5 lb/A)	102–168 kg P_2O_5/ha (91–150 P_2O_5 lb/A)	>168 kg P_2O_5/ha (>150 P_2O_5 lb/A)
P fertilizer application method (0.5)	None applied	Placed with planter deeper than 2 in.	Incorporated immediately before crop	Incorporated >3 months before crop or surface applied <3 months before crop.	Surface applied >3 months before crop
Organic P source application rate (1.0)	None applied	1–34 kg P_2O_5/ha (1-30 P_2O_5 lb/A)	35–67 kg P_2O_5/ha (31-60 P_2O_5 lb/A)	68–101 kg P_2O_5/ha (61-90 P_2O_5 lb/A)	>101 kg P_2O_5/ha (>90 P_2O_5 lb/A)
Organic P source application method (1.0)	None applied	Injected deeper than 2 in.	Incorporated immediately before crop	Incorporated >3 months before crop or surface applied <3 months before crop.	Surface applied to pasture or >3 months before crop

Notes: Site rating value for each parameter is multiplied by the corresponding weighting factor. Resulting values are summed to give the site P index. For irrigation erosion the QS is the flow rate in the furrow (Q) times the furrow slope (S).

4.2.2.1 Include STP Benchmarks Based on Environmental Risk and Not Crop Response

Long-term land application of organic residuals typically results in accumulation of P in surface (0-15 cm) soils beyond levels needed for optimum crop yields. This phenomenon is more acute for biosolids than manures because of the lower ratio of plant-available N to P in most biosolids. In developing nutrient management policies to control soil P loss, states attempt to define critical STP levels above which P delivery to water bodies is unacceptable. These so-called environmental STP thresholds are commonly based on readily available, agronomic soil P testing procedures. These testing procedures are, however, rarely calibrated for biosolids-amended soils.

One approach for establishing environmental thresholds is predicated on the rationale that soil P in excess of crop requirements is susceptible to release in runoff and drainage waters. This approach is simple to implement because traditional STP methods used for fertility recommendations are well-established in each state. Thus, some states (e.g., New Jersey, Illinois) directly use agronomic STP in their P index matrices and include labels like high or above optimum, which refer to crop response behavior. Descriptors of fertility status, however, may bear little direct relationship to offsite P migration potential.

A major drawback to equating agronomic and environmental thresholds is that the processes crops use to access soil P are different from those that determine susceptibility to solubilization by subsurface leaching or surface runoff (Kleinman et al., 2000). Plants can enhance P uptake by producing organic acids in the rhizosphere to solubilize water-insoluble P compounds. Thus, the soil P levels based on expected crop response cannot be directly translated into estimates of adverse water quality effects (Sharpley et al., 2003). There is a developing consensus among researchers that it is possible to maintain STP at levels that maximize crop yields while minimizing the risk of offsite P transport (Higgs et al., 2000).

As soil P concentrations increase, so does the potential for loss of P to surface and subsurface waters. There is a substantial body of research showing that the dissolved runoff Pis linearly correlated with the STP in the topsoil (Pote et al., 1999; McDowell and Sharpley, 2001; Andraski and Bundy, 2003). However, when a sufficiently wide range of STP levels is considered, the relationship becomes curvilinear (Figure 4-2) because of the increasing filling of P-fixing sites on the soil (McDowell and Sharpley, 2001). Surface soil testing is commonly used for crop fertilizer recommendations and most state P indices have STP as one predictor of the potential for environmental P loss. Although agronomic soil testing usually involves the 0–15 cm soil depth, the 0–5 cm depth is likely better correlated to for runoff P potential.

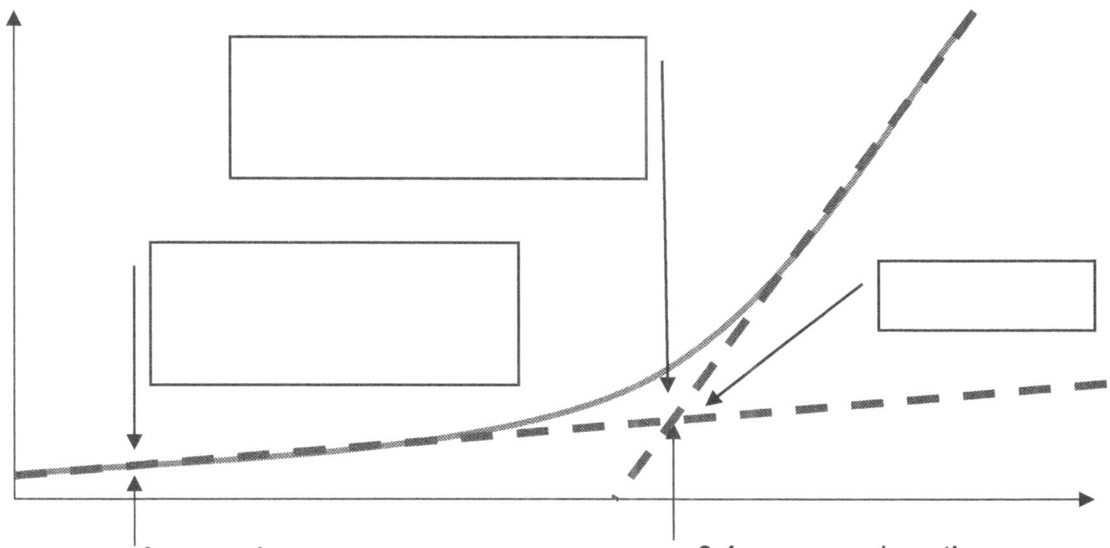

Figure 4-2. Illustration of Agronomic and Environmental Threshold Concepts.

While the curvilinear relationship between TRP and STP can be described by exponential models, a simple split-line model defines two linear sections with a change in slope occurring at the so-called "change point" (Figure 4-2). The change point ostensibly identifies the STP at which the soil P-sorbing sites become sufficiently filled that environmentally significant amounts of additional P will be mobilized by rainwater. Up to the change point, much of the P added is retained in the plow layer; beyond this point, P losses are substantially greater due to the reduced P-fixing ability of the soil.

In Pennsylvania, STP levels of 50 ppm P (Mehlich-3) are considered sufficient for crop production, and no additional fertilizer P would be recommended (Martz, 2002). However, the environmental loss threshold is considered to be 200 ppm STP (Sharpley et al., 2001). The state of Maryland has defined the environmental STP threshold as 75 mg kg^{-1} Mehlich-1 P, which is three times the agronomic critical level (Coale et al., 2002). Texas uses environmental soil test thresholds (200–500 ppm Bray-1 P, depending on site and watershed characteristics). Beyond the threshold no further P applications are allowed until changes occur that lower the site's P loss potential (e.g., reduction in STP). Some states have incorporated the environmental threshold principle into their P assessment policies, but many have not.

4.2.2.2 Recognize the Nature and Length of Buffer Zones in Protecting Surface Waters

The original P index (Table 4-1) was an "edge-of-field screening tool" that did not quantify the potential for transport from the field to the stream (Gburek et al., 2000). A site's vulnerability to P loss was evaluated without regard to the probability that surface runoff from the site would actually contribute to stream flow. Clearly, the proximity to a surface water body influences whether P lost from a site will enter the stream. For example, total P in runoff from poultry manure amended areas decreases with the length ratio of buffer zone to the waste-treated area (Bingham et al., 1980). Although the establishment of an effective setback distance is more important than maintaining a given ratio for large waste-amended areas, the data clearly show that total P levels in runoff decrease with buffer zone width.

As the P index concept evolved, the importance of evaluating a site's potential to impact water quality with reference to proximity to the receiving water body became apparent. Most state P indices now recognize the role of the buffer width (also called contributing distance or connectivity) in P loss potential assessment. Biosolids application sites have mandated field borders or setbacks distances. The 40 CFR Part 503 rules specify that biosolids shall not be applied to land "that is 10 meters or less from waters of the United States." Most states require 100- to 200-foot (30–60 m) buffer zones. State P management policies must recognize that these buffer zones reduce P loss to the receiving waterbody by slowing surface runoff and promoting infiltration and sediment trapping. The effect is more than retardation of P movement, as accumulated P can convert to insoluble forms.

Ideally, P assessment tools should reflect not only the distance from the edge of the field to the receiving waterbody but the *nature* of the intervening zone. Obviously, the potential for water quality impacts is quite different for a rock-lined channel and a riparian forest buffer. Management practices within the buffer zone (tillage, mineral fertilizer application, livestock grazing) also can have a major impact on P transport. Management policies should give "credit" for establishing control measures in the buffer zones that mitigate P loss from vulnerable biosolids application site. Dayton et al. (2003) demonstrated that application of WTRs to buffer strips at the edge of fields can remove soluble P from runoff waters. When repeated biosolids applications result in high STP, municipalities should have the option to "break the link" between the field and watercourse by establishing practices within the regulated buffer zones that reduce P transport.

4.2.2.3 Account for the Differences in P Loss Potential Among Biosolids Types

Biosolids have widely differing susceptibility to P solubilization by water (Brandt et al., 2004), and loss in surface runoff and subsurface drainage (Penn and Sims, 2002; Elliott et al., 2002; Sims et al., 2003; Elliott et al., 2005a). Figure 4-3 shows the dramatic difference in P solubility among fertilizer (TSP), manures, and biosolids; conventionally treated biosolids exhibiting significantly lower P solubility.

Figure 4-4 shows the total dissolved P following simulated rainfall for 14 treatments (10 biosolids, three manures, and one control) as a function of the total P loading rate (kg P ha^{-1}). There was no positive correlation between the P concentration in runoff and the biosolids-P loading rate. This dramatically illustrates that P_T loading is an unreliable guide to the amount of biosolids-P lost in runoff. Indeed the greatest effective P loading (555 kg ha^{-1} for a biosolids treated by advanced alkaline stabilization) resulted in the smallest total dissolved P value (<0.1 mg L^{-1}). There were seven biosolids types with similar P_T loadings (~180-220 kg ha^{-1}), but the mean total dissolved P values were from <1 mg L^{-1} to >20 mg L^{-1}.

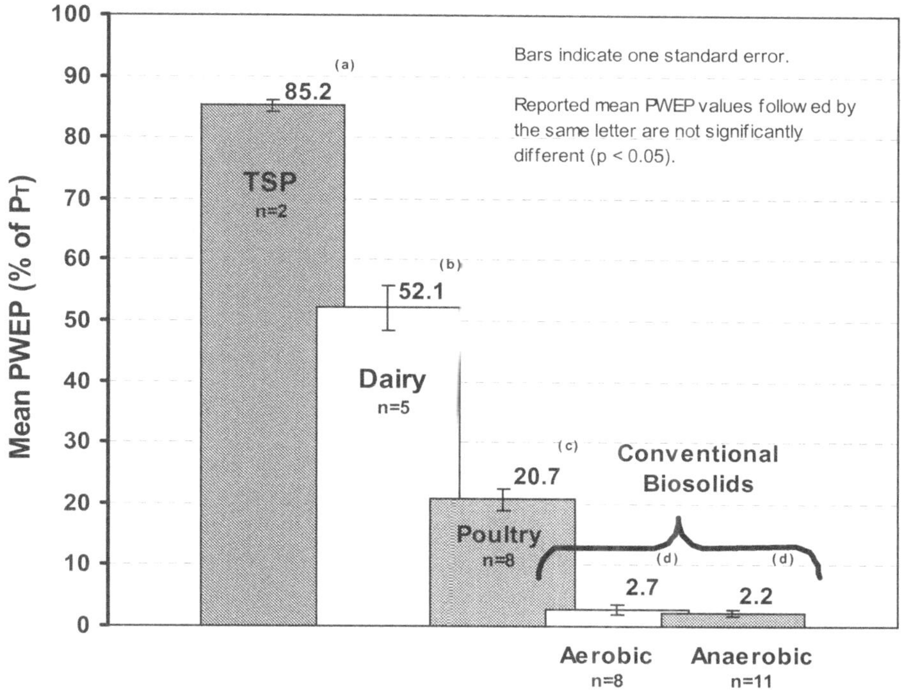

Figure 4-3. Comparison of Percent Water-Extractable P (PWEP) for TSP, Manures, and Conventionally Treated Biosolids. Reprinted with permission from Brandt et al., 2004. Copyright 2004, Water Environment Federation, Alexandria, Va.

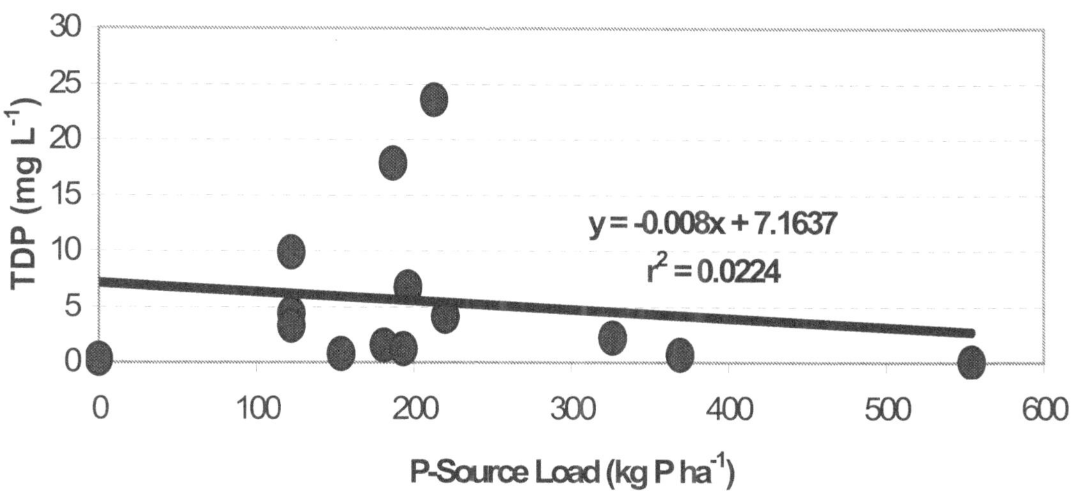

Figure 4-4. Mean Total Dissolved P Concentrations versus P-Source Loadings.

To account for such observations that P release to runoff varies widely for different materials, some site assessment tools allow for differential weighting of applied P sources. The weighting factor that differentiates P sources based on their relative potential to release P into runoff is the P source coefficient, or PSC (Leytem et al., 2004). In the P index calculation, the PSC of an amendment is multiplied by amendment total P application rate, with the product viewed as the applied P susceptible to offsite transport. Thus, a PSC, as the fraction of the total applied P available for transport, can take on any value between 0.2 and 1.0 (Table 4-2). Because the water-extractable P content of the applied material is correlated with dissolved P in runoff (Kleinman et al., 2002b), water-extractable P is seen as a key means of developing PSCs in the P index (Leytem et al., 2004).

If the P-source loads shown in Figure 4-4 are multiplied by source-appropriate PSC values (Table 4-2), the product reflects the amount of applied P susceptible to runoff. The runoff data of Figure 4-4 are plotted versus this modified P source loading rate in Figure 4-5a. The positive slope is consistent with the expectation that total dissolved P should increase with the amount of runoff-soluble applied P. The regression relationship however, is weak ($r^2 = 0.45$) because the tabulated PSC values are single descriptors that do not account for all the factors influencing P loss potential.

Table 4-2. Organic PSCs in the Pennsylvania P Index. Reprinted with permission from Weld et al., 2003. Copyright 2003, The Pennsylvania State University, University Park, Pa.

P Source	PSC
Swine slurry	1.0
Poultry	
Broiler	0.8
Layer	0.9
Turkey	0.9
Duck	0.9
Dairy	
Liquid	0.9
Bedded pack	0.8
Beef	0.8
Alum treated manures	0.5
Biosolids	
BPR	0.8
Alkaline stabilized	0.4
Conventionally stabilized	0.3
Composted	0.3
Heat-dried	0.2
Advanced alkaline stabilized	0.2

Because it is not realistic to provide discrete PSC values for all types of manures, biosolids, composts, and other land applied P sources, the predictive capability of P site indices should be improved if PSCs could vary continuously to describe amendments with the entire spectrum of P solubilities. Thus, efforts are underway to develop source-specific PSCs based on the measured water-extractable P of the applied material. Based on numerous rainfall-runoff studies that followed the National Phosphorus Research Project protocols, Elliott et al. (2006) developed an empirical equation for computing source-specific PSCs from laboratory-determined water-extractable P values [$PSC = 0.102 \times \text{water-extractable } P^{0.99}$]. When the P loading rates were multiplied by the PSC computed by this expression, a much stronger relationship ($r^2 = 0.86$) was observed (Figure 4-5b).

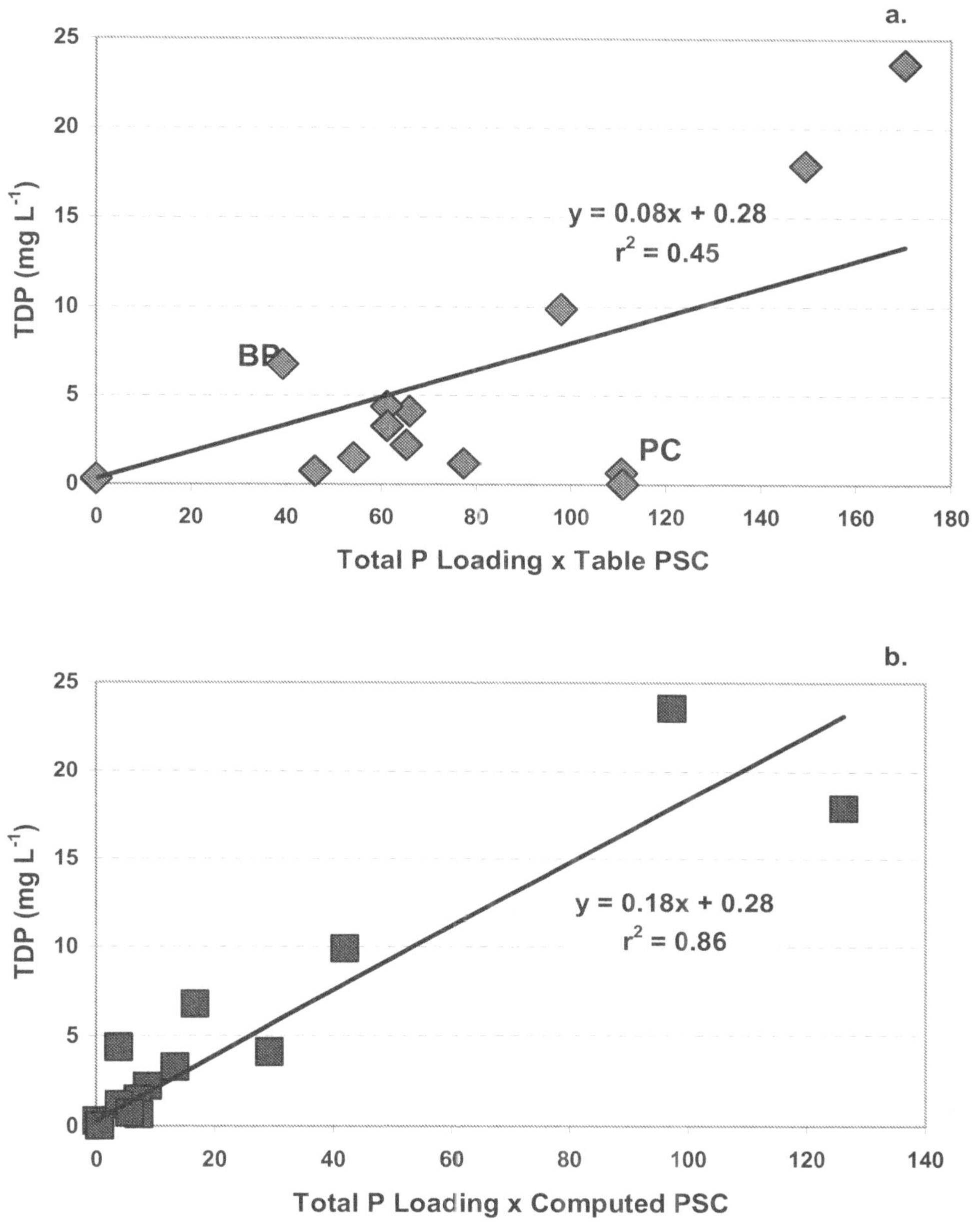

Figure 4-5. Mean Total Dissolved P versus Soluble Applied P Determined Using (a) Table 4-2 PSCs and (b) Computed PSCs. PC, Philadelphia cake; BP, Baltimore Patapsco biosolids.

Examination of individual data points reveals at least two situations where the computed PSC are clearly superior to default table values. Some biosolids, like Philadelphia cake (labeled PC in Figure 4-5a), have very high Fe or Al content because of the influent composition or intentional additions of salts during wastewater treatment. Although the Pennsylvania P index has a unique PSC value for alum-treated manures (Table 4-2), there is no analogous PSC for biosolids with high Al and/or Fe. Another problem with tabulated PSCs is the confusion arising when a material appears to fit in more than one category. The Baltimore Patapsco biosolids (labeled BP in Figure 4-5a) is a heat-dried product (PSC = 0.2) generated in a BPR wastewater treatment process (PSC = 0.8). Based on a relatively low water-extractable P, the researchers selected PSC of 0.2 for this biosolids material in constructing Figure 4-5a. Source coefficients based on testing of the actual materials eliminates this confusion of assigning P sources to a unique category.

Source-specific PSCs should enhance the ability of assessment tools to identify vulnerable sites. However, of the 41 state P indices evaluated by Sharpley et al. (2003), only nine states (Arkansas, Delaware, Florida, Georgia, Louisiana, Maryland, Pennsylvania, Tennessee, and Virginia) include an estimate of the availability or solubility of the P applied to a site. Of these states, not all make specific reference to biosolids or, if they do, a "one-size-fits-all" approach has been used, which considers biosolids equal to manures. Unless the variable and, typically, lower solubility of P in biosolids is taken into account, the environmental impact of biosolids application will likely be overestimated by P site index assessment tools.

Another reason for including source-specific PSC values in P indices is to provide a management tool for reducing P flux from biosolids-treated soils. The potential for offsite P migration at land application sites strongly depends on biosolids composition and processing mode. Thus, amending biosolids with metal salts or coagulation water treatment residuals will reduce P runoff and leaching potential (see Section 3.3.3). Without PSC values, a P site assessment tool will not reflect such P loss reduction management practices.

4.2.2.4 Recognize the Variable P Fertilizer Replacement Value of Different Biosolids Types

Application rates may be limited to crop removal of P when the site rating dictates P-based nutrient management. For manures, state agronomy guide removal rates can simply be matched to total manure P content because manure P is assumed equal to fertilizer P for supplying crop needs and building soil P levels (Martz, 2002). Because biosolids are P-rich relative to manures, assuming 100% plant availability of P will restrict biosolids application rates to 1–2 Mg ha^{-1} (~0.5–1 ton ac^{-1}) because most crops only remove 10–40 kg P ha^{-1} (O'Connor et al., 2004). Such low application rates will severely reduce a farmer's interest in utilizing biosolids because supplemental fertilizer N will be necessary and low rates are uneconomical and impractical. Low application rates will also mean municipalities will need much greater acreage to dispose of their biosolids.

The P-supplying ability of most biosolids is, however, generally less than commercial P fertilizer. Greenhouse studies were conducted with a common pasture grass grown in two P-deficient soils amended with 12 biosolids and a commercial fertilizer (TSP) to quantify P uptake and to assess the relative phytoavailabilities of the P-sources (O'Connor et al., 2004). The mean phytoavailabilities relative to TSP are shown in Figure 4-6. Biosolids were grouped into three general phytoavailability categories relative to TSP: high (>75% of TSP), moderate (25–75% of TSP), and low (<25% of TSP). Two biosolids, produced via BPR processes and conventionally stabilized (i.e., aerobic or anaerobic digestion), were in the high category and mimicked fertilizer-P with regard to P phytoavailability. Most biosolids produced by conventional wastewater treatment and.

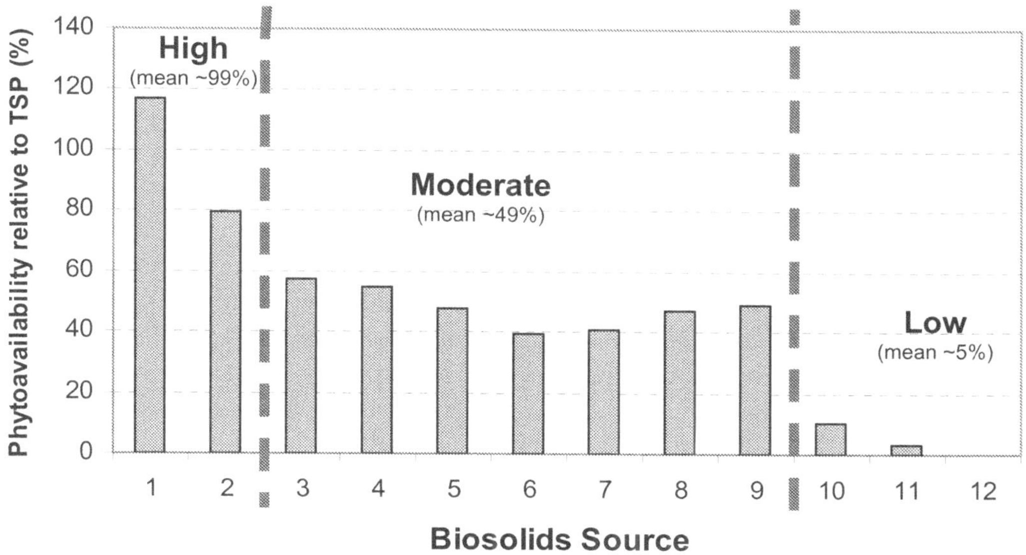

Figure 4-6. Mean Phytoavailability of Biosolids-P Relative to TSP Mineral Fertilizer. Proposed three categories: availability comparable to mineral fertilizer for BPR biosolids (1-2), moderate availability for most biosolids (3-9), and low availability for High-Fe or -Al heat-dried (10-11) and advanced alkaline stabilized biosolids (12). Reprinted with permission from O'Connor et al., 2004. Copyright 2004, Water Environment Research Foundation, Alexandria, Va.

solids digestion and additional treatments (e.g., composting) were in the moderate category. Also included in this category were a BPR that had been pelletized and a BPR supplemented with Al. The low category included biosolids treated by advanced alkaline stabilization (e.g., N-Viro process) and biosolids containing greater than normal (>50 g kg^{-1}) total Fe and Al concentrations *and* processed to high (>60%) solids content.

The U.S. EPA Process Design Manual for Land Application (U.S. EPA, 1995) suggests a single 50% relative effectiveness factor for biosolids-P. If a state P index assessment for a site dictates P-based application, the fraction of total P available to crops must be known. Although the 50% value may be a reasonable estimate for many biosolids (e.g., medium category in Figure 4-6), some are clearly much more, and some much less, able to supply crop P needs.

Agronomically, the differences in relative phytoavailability of various biosolids (relative to fertilizer-P) are most obvious in P-responsive (deficient) soils. Thus, very sandy soils with low P-sorbing capacities are expected to reflect biosolids-P differences best. Even sandy soils, however, that test high in STP and heavier textured soils with appreciable P-sorbing capacity often fail to reflect biosolids-P differences. In such soils, additions of even highly soluble fertilizer-P sources are of limited agronomic benefit, and differences in biosolids-P are unremarkable. Phosphorus solubility and phytoavailability in such soils is controlled by the soil, whereas in poorly P-sorbing soils P chemistry is controlled by biosolids-P characteristics. The researchers' data suggest little justification for distinguishing between biosolids source when application rates on strongly P-sorbing soils are dictated to be P-based.

Many soils in the United States and Europe test high to very high in STP because of historical fertilizer and/or waste management practices. Further additions of P (as fertilizer or biosolids) have little agronomic benefit, but biosolids additions as a disposal mechanism can still be guided by differences in P solubility when environmental P issues are paramount. Biosolids-P solubility (well estimated by water-extractable P values) determines P phytoavailability in P-responsive soils. The same characteristic controls P leaching and soluble P runoff. Thus, highly soluble biosolids-P sources (e.g., BPR materials low in total Fe and Al) can be expected to mimic fertilizer-P behavior. Very low solubility biosolids-P sources (heat dried, high total Fe and Al) can have reduced agronomic and environmental lability. Short-term greenhouse and rainfall simulation studies and the much rarer field studies suggest that low solubility biosolids-P sources can be applied at N-based rates without endangering crops or water quality. Additional, long-term field research is critical to judging the agronomic and environmental safety of the practice. Validation would foster the sustainability of biosolids land application programs.

4.2.2.5 Standard Tests for Characterizing the Mobility and Bioavailability of Biosolids P

Environmental effects are paramount in addressing P in land-based recycling of biosolids, and agronomic P source and soil characterization parameters may be inappropriate system descriptors. States have existing protocols for verifying the P content of mineral P fertilizers. In Pennsylvania, for example, the neutral ammonium citrate extraction (Kane, 1995) is the official method for determining guaranteed "available" P (i.e., P available for plant uptake) in commercial fertilizers. Some states, like Maine (Maine Department of Environmental Protection, 1999), require that plant-available P be determined for land-applied biosolids. However, Elliott et al. (2005b) tested 35 biosolids and found that neutral ammonium citrate-P was not different (P = 0.05) from total P using strong acid digestion by the EPA 3051 method. Moreover, P phytoavailability to pasture grass was not correlated ($r^2 = 0.10$) with biosolids neutral ammonium citrate-P. The greenhouse data of Phase I confirmed that biosolids total P and common soil test-P methods are unreliable indicators of the RPP of biosolids.

A major conclusion of this work is that the biosolids water-extractable P is a useful parameter for assessing the RPP (Table 3-1) and environmental loss potential (Figure 3-10) of land-applied biosolids. For water-extractable P to be useful as an indicator of P loss potential and RPP across materials of varying properties, a common protocol needs to be adopted for measuring water-extractable P in biosolids, manures, and other land-applied P sources. Recently, Kleinman (2006) led a major study to evaluate the water-extractable P test for manures and biosolids. Based on multilaboratory testing and runoff studies, a protocol for measuring water-extractable P was recommended. The protocol selected was consistent with that used in this study except that a 1:100 (dry solids-to-water) ratio was selected. The researchers employed a 1:200 extraction ratio (see Section 2.2.1), which had been developed and published by Wolf et al. (2005). To assess the impact of extraction ratio on biosolids water-extractable P values, 10 biosolids of varying properties and from different processes were evaluated at the 1:100 and 1:200 extraction ratios (Appendix Table A-9). The 1:200 ratio gives water-extractable P values that are 20–30% higher than values at the 1:100 ratio. There are several compelling reasons to prefer the 1:200 ratio for biosolids. The 1:200 ratio provides better discrimination among P biosolids types; it can be theoretically used with materials of solids contents as low as 0.5% (compared to 1% for the 1:100 ratio), and a tremendous amount of correlation work with biosolids has already been done using the 1:200 ratio. Moreover, the state of Wisconsin requires municipalities to report biosolids water-extractable P based on the 1:200 ratio test. Despite the stated advantages of the 1:200 protocol, if the 1:100 method is adopted nationally (and states use the water-extractable P as part of P-based nutrient management policies), municipalities will be compelled to follow the recommended testing procedure. This is currently an unresolved issue.

4.3 Research Needs

Phases I and II of the research provided background information on the environmental behavior and agronomic benefit of various biosolids products. As P management policies have been developed simultaneously with this research, the researchers have identified several unresolved issues that must be addressed for the continuance of land application as a sustainable biosolids management strategy. The researchers are convinced that P-based nutrient management represents a substantial challenge for biosolids recycling programs in many states.

4.3.1 Strategies for Recycling BPR Biosolids and Other High P-Solubility Products

Research conducted in Phase I and II documented the exceptionally high P-loss potential and phytoavailability of BPR biosolids compared to other products. Additional work with BPR materials appears to be especially critical because of the growing trend toward replacing chemical-intensive treatment approaches with BPR processes in new construction and modifications of existing WWTPs (Daigger, 2003). Impacts of treatments to modify the high solubility of BPRs via metal additions, heat drying, or composting on the long-term behavior of residual-P should be quantified. WTRs have been shown to reduce P solubility of BPR materials and, hence, agronomic and environmental lability. Incorporation and thorough mixing of BPR materials in most soils reduces environmental lability (reduces runoff). Field studies with BPR materials are scarce, and critical long-term studies are non-existent. The solubility of P generated by emerging technologies (e.g., membrane bioreactors, denitrification filters) also needs to be evaluated.

4.3.2 Evaluate the Long-term Stability of Low P-Solubility Residuals

The long-term stability of low P solubility residuals remains an unresolved issue. This is particularly true for organic nutrient sources where the low loss potential of P is attributable to the formation of Al- or Fe-bound P compounds. If P is precipitated as, or adsorbed to, Fe(III) phases, release of P under anaerobic conditions (e.g., buried in aquatic sediments) is possible. Necessary studies would include laboratory characterization of materials incubated for long periods and/or materials from actual field application studies. Residuals-P presence in various water sources should be related to actual biological availability via chemical determinations and bioassays, as needed. Different environmental conditions (pH, redox, etc.) could be imposed to identify critical conditions that impact ultimate P solubility.

4.3.3 Evaluate the Role of Buffer Strip Characteristics in Mitigating P Loss

The important role of buffer strips in protecting surface water quality was noted earlier in this chapter. Particularly lacking are actual field measurements focused on the role of buffer zone characteristics on runoff collection and characterization. Since state regulations for biosolids application sites have mandated field borders or setbacks distances, this is particularly important for biosolids recycling. P assessment tools should reflect not only the distance from the edge of the field to the receiving water body but the *nature* of the intervening zone. Some state P indices address the benefits of riparian buffer strips, but this component of P indices needs more refinement and is especially important to biosolids recycling. Recently, Dayton and Basta (2005) showed that buffer strips amended with WTRs provide additional P transport reduction potential, but most P indices do not consider this as a best management practice to lower a site's P risk index score. If states are to give credits for establishing control measures in the buffer zones that mitigate P loss from vulnerable biosolids application sites, such policies must be based on sound scientific research.

APPENDIX A

A-1. Immokalee Soil General Chemical Properties Measured in 2001 from Kirton Ranch Soil Samples.

	A-horizon (0-10cm)	E -horizon (~ 20 -30 cm)	B_h -horizon (~40-50 cm)
pH	5.5	5.9	5.1
C (g kg^{-1})	12 ±0.1	3.0 ±0.0	17 ±0.2
Total P (mg kg^{-1})	24.5 ± 5.46†	7.9 ± 4.56	24.5 ± 11.3
Total Al (mg kg^{-1})	106 ± 23.3	33.6 ± 7.55	1280 ± 770
Total Fe (mg kg^{-1})	101 ± 39.6	38.5 ± 6.21	94.8 ± 23.2
Total Ca (mg kg^{-1})	449 ±7.6	134 ±1.1	228 ±7.3
Total Mg (mg kg^{-1})	36 ±3.2	16 ±0.1	75 ±0.9
Oxalate†† P (mg kg^{-1})	10.0 ± 2.95	3.76 ± 2.76	23.8 ± 13.7
Oxalate Al (mg kg^{-1})	54.8 ± 6.74	15.8 ± 5.35	970 ± 418
Oxalate Fe (mg kg^{-1})	61.7 ± 8.52	13.0 ± 5.98	39.0 ± 5.08
Mehlich 1-P (mg kg^{-1})	7.03 ± 6.54	1.88 ± 1.33	12.1 ± 12.8
Water-extractable P (mg kg^{-1}) †††	2.88 ±0.19		
KCl-P (mg kg^{-1})	3.87 ± 0.78	0.77 ± 0.32	3.31 ± 0.52
NaOH-P (mg kg^{-1})	9.87 ± 1.23	5.49 ± 1.02	21.3 ± 3.2
HCl-P (mg kg^{-1})	3.78 ± 1.02	3.14 ± 0.88	4.56 ± 1.10
ISP (mg kg^{-1})	9.6 ± 0.11	6.3 ± 0.12	16.4 ± 0.25
DPS ††††	0.10 ± 0.09	0.14 ± 0.15	0.02 ± 0.01

† Means of three samples ± standard deviation
†† 0.2 M oxalate
††† Soil: solution = 1: 10
†††† DPS = degree of phosphorus saturation [oxalate P / oxalate Fe + oxalate Al (in moles)]

A-2. General Chemical Properties of SFWMD Amendments (Oven Dry Basis).

SOURCE	Chicken manure	Boca Raton Biosolids	Pompano Biosolids	TSP	Al-WTR
pH	7.7	7.6	7.6	5.9	5.5
C (%)	32.0	34.7	36.6	-	19.9
N (%)	8.4	50.4	43.1	-	0.7
% Solids	27.0 ± 0.3†	13.4 ± 0.04	15.4 ± 0.04	100	62.5 ± 2.2
Total P (g kg^{-1})	25.3 ± 1.1	47.3 ± 2.3	26.2 ± 0.2	209	4.6 ± 0.7
Total Al (g kg^{-1})	0.9 ± 0.07	9.3 ± 0.4	9.2 ± 0.4	10.0	157.3 ± 3.4
Total Fe (g kg^{-1})	1.5 ± 0.1	24.3 ± 0.8	32.8 ± 0.4	15.7	6.0 ± 0.1
Total Ca (g kg^{-1})	102 ± 3.2	27.5 ± 1.1	47.0 ± 0.5	137	1.5 ± 0.1
Total Mg (g kg^{-1})	5.8 ± 0.2	10.0 ± 0.5	4.1 ± 0.1	6.2	0.4 ± 0.02
Oxalate†† P (g kg^{-1})	12.7 ± 0.0	34.0 ± 0.9	20.4 ± 0.1	186	4.3 ±0.0
Oxalate Al (g kg^{-1})	0.2 ± 0.0	8.9 ± 0.6	9.2 ± 0.0	6.9	145 ±0.4
Oxalate Fe (g kg^{-1})	0.7 ± 0.0	19.4 ± 0.5	24.7 ± 0.2	11.0	5.1 ±0.0
Oxalate Ca (g kg^{-1})	-	-	-	-	0.33 ±0.01
Oxalate Mg (g kg^{-1})	4.2 ± 0.1	9.7 ± 0.0	3.7 ± 0.2	-	0.36 ±0.01
P Saturation Index †††	21.7 ± 0.56	1.44 ± 0.02	0.7 ± 0.02	-	0.02 ± 0.0
Water-extractable P (g kg^{-1}) ††††	4.57± 0.16	5.52 ±0.18	1.16 ±0.08	175	0.03 ±0.00
Percent Water-extractable P	18.1	11.7	4.43	83.7	-

† Means of three samples ± standard deviation
†† 0.2 M oxalate
††† Phosphorus Saturation Index [oxalate P / oxalate Fe + oxalate Al (in moles)]
†††† Soil : Solution 1:200

A-3. Plant Dry Matter Yields, Weighted P Concentrations, and P Uptake from the SFWMD Study.

Treatments	<----2003 Harvest--->			<----2004 Harvest--->		
	Dry matter	Weighted P conc.	P uptake	Dry matter	Weighted P conc.	P uptake
	Mg ha^{-1} soil	g kg^{-1}	kg P ha^{-1}	Mg ha^{-1} soil	g kg^{-1}	kg P ha^{-1}
Controls	5.8 ±1.27	2.28 ± 0.11	13.4 3.58	5.85 ±0.63	2.44 ± 0.23	14.2 ± 0.9
Manure-N, no WTR	5.30 ±1.82	2.43 ± 0.48	12.60 ±2.82	6.05 ±0.95	2.02 ± 0.33	12.4 ± 3.7
Manure-N, WTR	4.78 ±1.40	2.17 ± 0.37	10.7 ±4.9	5.78 ±0.63	1.69 ± 0.19	9.70 ± 0.83
Manure-P, no WTR	5.20 ±0.44	1.93 ± 0.05	10.0 ±1.07	6.96 ±0.62	2.10 ± 0.46	14.8 ± 4.7
Manure-P, WTR	6.68 ±2.85	1.89 ± 0.55	13.6 ±9.9	5.73 ±0.62	1.71 ± 0.28	9.92 ± 2.65
Boca-N, no WTR	6.60 ±0.60	2.76 ± 0.27	18.3 ±2.1	5.63 ±0.45	2.31 ± 0.12	13.0 ± 0.4
Boca-N, WTR	6.43 ±2.88	2.04 ± 0.19	12.7 ±4.4	6.04 ±1.29	1.86 ± 0.36	10.9 ± 0.6
Boca-P, no WTR	6.90 ±0.15	2.60 ± 0.09	18.0 ±0.61	6.63 ±0.72	2.77 ± 0.21	18.5 ± 3.4
Boca-P, WTR	5.37 ±1.40	2.03 ± 0.10	11.0 ±3.5	6.29 ±0.68	1.50 ± 0.12	9.39 ± 0.43
Pompano-N, no WTR	7.70 ±1.84	2.70 ± 0.69	20.7 ±7.10	6.14 ±0.82	2.66 ± 0.11	16.3 ± 1.6
Pompano-N, WTR	6.39 ±1.66	2.38 ± 0.10	15.3 ±4.6	6.50 ±0.43	1.54 ± 0.15	10.0 ± 1.5
Pompano-P, no WTR	5.80 ±0.98	2.14 ± 0.45	12.1 ±0.65	6.03 ±0.84	2.40 ± 0.08	14.5 ± 1.8
Pompano-P, WTR	5.51 ±0.20	1.78 ± 0.09	9.80 ±0.83	5.49 ±0.62	1.84 ± 0.17	10.1 ± 0.8
TSP-N, no WTR	6.40 ±0.32	2.56 ± 0.25	16.4 ±1.97	6.78 ±0.58	2.51 ± 0.12	17.1 ± 2.1
TSP-N, WTR	5.79 ±1.66	2.06 ± 0.43	12.2 ±5.2	6.17 ±0.38	1.98 ± 0.24	12.3 ± 2.2
TSP-P, no WTR	6.20 ±1.22	2.64 ± 0.29	16.2 ±3.28	5.75 ±0.25	2.27 ± 0.18	13.1 ± 1.4
TSP-P, WTR	5.96 ±1.14	1.93 ± 0.30	11.6 ±3.1	5.93 ±0.97	1.71 ± 0.52	9.82 ± 1.40

A-4. ANOVA Results for Soil Test P and Plant Data, SFWMD Study.

Source of variation	Water-Extractable P			ISP			M-1P			Total P			2003			2004		
	Jun '03	Jan '04	Mar '04	Jun '03	Jan '04	Mar '04	Jun '03	Jan '04	Mar '04	Jun '03	Jan '04	Mar '04	Dry Matter	P conc.	P uptake	Dry Matter	P conc.	P uptake
Source of P (S)	NS	NS	NS	*	NS	NS	NS	**	NS	*	**	**	NS	NS	NS	NS	*	NS
Appl. rate of P source (R)	NS	NS	NS	NS	NS	NS	NS	**	NS	**	*	**	NS	**	NS	NS	NS	NS
Appl. rate of WTR (W)	**	**	**	**	**	**	NS	*	**	**	**	**	NS	**	**	NS	**	**
S X R	NS	NS	NS	NS	NS	NS	NS	*	**	NS	*	**	NS	NS	NS	NS	NS	**
S X W	NS	NS	NS	NS	NS	NS	NS	*	NS	NS	**	NS	NS	NS	NS	NS	*	NS
R X W	NS	NS	NS	NS	NS	NS	NS	*	NS	NS	NS	*	NS	NS	NS	NS	NS	NS
S X R X W	NS	NS	NS	NS	NS	NS	NS	*	*	NS	NS	NS	NS	NS	NS	NS	**	*

* means significant at 5%
** means significant at 1%
NS means not significant at 5%

A-5. Correlation Coefficients between Various Shallow (0–5 cm) Soil P Measures and Plant P Uptake, SFWMD Study.

Soil P	With WTR Data			Without WTR Data		
	2003	2004	Total	2003	2004	Total
Water-Extractable P	0.44881†	0.6238	0.4833	0.396	0.03911	0.71496
	0.001‡	<.0001	<.0001	0.0409	0.8464	<.0001
ISP	0.42026	0.2863	0.40702	0.40093	-0.03092	0.36596
	0.0021	0.041	<.0001	0.0382	0.8783	0.0065
M1P	0.0683	0.1026	0.41117	0.3306	0.24736	0.44262
	0.6339	0.473	<.0001	0.0921	0.2135	0.0008
Total P	-0.13727	-0.3962	0.19689	0.45792	0.39645	0.64812
	0.3575	0.004	0.052	0.0213	0.0406	<.0001
0.2M Oxalate P	-0.08494	-0.4305	0.0985	0.37161	0.0007	0.19709
	0.5534	0.001	0.3247	0.0563	0.9972	0.1532

† Correlation coefficient (*r*)

‡ *P* value

A-6. Degree of P Saturation of Soil Samples from the SFWMD Study.

Treatments	Jun '03 (0–5cm)	Jan '04 (0–5cm)	Dec '04 (0–5cm)	Mar '04 (0–15cm)	Dec '04 (0–15cm)
Controls	61.8	52.7	24.0	51.8	79.2
Manure-N, no WTR	147	113	146	47.1	57.1
Manure-N, WTR	21.8	14.5	26.2	14.5	13.7
Manure-P, no WTR	65.5	58.2	91.1	47.3	61.3
Manure-P, WTR	18.2	14.5	15.7	11.1	14.2
Boca-N, no WTR	69.1	65.5	92.0	32.9	92.6
Boca-N, WTR	27.3	18.2	46.0	9.05	12.9
Boca-P, no WTR	81.8	65.5	75.3	36.6	81.6
Boca-P, WTR	25.5	18.2	28.6	9.71	14.0
Pompano-N, no WTR	102	85.5	55.6	44.6	76.5
Pompano-N, WTR	14.5	12.7	29.2	6.42	5.65
Pompano-P, no WTR	65.5	60.0	51.4	37.3	71.6
Pompano-P, WTR	14.5	10.9	23.1	4.23	6.67
TSP-N, no WTR	155	90.9	170	49.1	71.9
TSP-N, WTR	21.8	14.5	33.7	12.8	19.2
TSP-P, no WTR	144	58.2	159	36.2	73.0
TSP-P, WTR	18.2	20.0	29.2	17.1	18.4

A-7. Masses of P Forms Lost in Runoff and Leachates, WTR Amendment Simulations. (Flow-weighted Averages of Three Replicates).

P source	Rate	Mass of P applied	WTR	Soil P load mass[Ω]	<---- Runoff P					---->	<---- Leachate P (mg)			---->	Total P loss	[Φ]Total BAP loss	% of applied P lost as BAP
					TP†	SRP‡	TDP*	DOP¥	PP#	BAP@	TP†	SRP‡	PP#	BAP@			
	kg ha^{-1}	mg	%	mg													%
TSP	224	1680	0	2048	27.3	11.8	13.0	1.20	14.4	11.9	746	706	40.2	536	774	548	32.6
TSP	56	420	0	788	41.7	15.9	17.6	1.64	24.1	23.1	147	134	13.8	133	189	156	37.1
Manure	224	1680	0	2048	492	96.0	104	7.94	388	146	66.3	63.1	3.18	43.3	558	190	11.3
Manure	56	420	0	788	91.5	18.7	20.0	1.38	71.5	48.4	83.8	65.5	18.2	64.1	175	113	26.9
Boca	224	1680	0	2048	391	45.3	47.4	2.14	343	88.6	126	114	12.1	97.0	516	186	11.1
Boca	56	420	0	788	65.9	7.96	8.52	0.56	57.3	22.1	35.7	30.8	4.92	31.2	193	53.3	12.7
Pompano	224	1680	0	2048	111	5.98	6.82	0.84	104	23.5	40.1	34.6	5.45	39.3	151	62.8	3.74
Pompano	56	420	0	788	77.7	2.24	2.65	0.41	75.0	11.3	46.5	34.3	12.2	33.1	124	44.3	10.5
Control	0	0	0	368	8.97	0.99	1.16	0.17	7.8	3.30	29.2	25.1	4.02	26.1	38.1	29.5	-
TSP	224	2370	1	2738	33.8	12.9	13.4	0.50	20.5	15.4	620	524	95.2	448	653	464	19.6
TSP	56	1110	1	1478	15.6	4.42	5.2	0.74	10.4	5.20	171	162	8.18	140	186	146	13.2
Manure	224	2370	1	2738	453	47.2	58.2	11.0	395	90.3	51.4	38.8	12.5	23.6	505	114	4.81
Manure	56	1110	1	1478	48.9	5.80	6.06	0.27	42.8	12.7	51.1	30.8	20.4	34.3	100	47	4.23
Boca	224	2370	1	2738	157	22.6	24.3	1.74	132	35.2	35.7	30.8	4.92	28.9	192	64.2	2.71
Boca	56	1110	1	1478	79.6	6.69	7.06	0.36	72.6	15.1	35.7	23.9	11.9	30.3	115	45.4	4.09
Pompano	224	2370	1	2738	137	4.16	4.94	0.78	132	15.5	22.9	21.1	1.80	15.9	160	31.5	1.33
Pompano	56	1110	1	1478	41.4	1.38	3.17	1.79	38.2	4.67	23.0	21.0	2.08	19.1	64.4	23.8	2.14
Control	0	690	1	1058	9.10	0.38	1.24	0.86	7.9	0.91	20.2	18.5	1.66	16.4	29.3	17.3	2.51

Ω Calculated as P added (source) + soil P (control) + WTR P (when applied)

† TP = Total P. ‡ SRP = Soluble reactive P. *TDP = Total dissolved P. ¥DOP = Dissolved organic P. #PP = Particulate P. @BAP = Bioavailable P (ISP)

Φ Total BAP loss is sum of runoff and leachate BAP mass loss)

* Significant difference between with and without WTR at $P < 0.05$ by contrast

*** Significant difference between with and without WTR at $P < 0.0001$ by contrast

NS not significant difference between with and without WTR at $P = 0.05$ by contrast

A-8. ANOVA Table for Varying Forms of P Loss in the WTR Simulation Study.

Source of variation	BAP weighted P conc. (mg kg^{-1})		BAP (mg)			Runoff P loss (mg)					Leachate P loss (mg)			Total P loss (mg)	% P applied lost as BAP
	Leachate	Runoff	L-P loss	R-P loss	Total loss	TP loss	SRP loss	TDP loss	PP loss	DOP loss	SRP loss	TP loss	PP loss		
Source	***	***	***	***	***	***	***	***	***	**	***	***	NS	***	***
rate	***	***	***	***	***	***	***	***	***	*	***	***	NS	***	NS
WTR	***	***	NS	**	*	NS	***	**	NS	NS	*	*	NS	*	***
Source*rate	***	**	***	***	***	***	***	***	***	**	***	***	NS	***	NS
Source*WTR	***	***	NS	*	NS	NS	***	**	NS	NS	NS	NS	NS	NS	*
rate*WTR	*	NS	NS	NS	NS	NS	*	*	NS	NS	*	NS	NS	NS	*
Source*rate*WTR	NS	NS	NS	NS	NS	*	*	*	*	NS	*	NS	NS	*	NS

[*] Significant difference between with and without WTR at $P < 0.05$ by contrast

[***] Significant difference between with and without WTR at $P < 0.0001$ by contrast

[NS] not significant difference between with and without WTR at $P = 0.05$ by contrast

TP = Total P; SRP = Soluble reactive P; TDP = Total dissolved P; DOP = Dissolved organic P; PP = Particulate P; BAP = Bioavailable P (ISP)

A-9. Comparison of Biosolids Water-Extractable P at 1:100 and 1:200 Extraction Ratios

Facility Designation	Water-Extractable P (g kg^{-1}) 1:100 Extraction Ratio*	Water-Extractable P (g kg^{-1}) 1:200 Extraction Ratio*
A	0.195	0.174
B	0.076	0.120
C	0.109	0.139
D	0.061	0.060
E	0.012	0.013
F	0.126	0.137
G	0.239	0.349
H	0.609	0.849
I	0.703	0.854
J	1.075	1.175

*Each value is mean of four samples with P determination by ICP analysis.

References

Andraski, T.W.; Bundy, L.G. Relationship between Phosphorus Levels in Soil and in Runoff from Corn Production Systems. *J. Environ. Qual*. **2003**, *32*, 310–316.

Beegle, D.R. Soil Fertility Management. In *The Agronomy Guide*, Martz, E., Ed.; Publication Distribution Center, Pennsylvania State University: University Park, Pa., 2002; pp 19–42.

Bingham, S.C.; Westerman, P.W.; Overcash, M.R.. Effect of Grass Buffer Zone Length in Reducing the Pollution from Land Application Areas. *Trans. ASAE* **1980**, *23*, 330–342.

Brandt, R.C. Land application of biosolids under phosphorus-based nutrient management. PhD. Dissertation, Pennsylvania State University: University Park, Pa., 2003.

Brandt, R.C.; Elliott, H.A. Phosphorus Runoff Losses from Surface-Applied Biosolids and Dairy Manure, *Proceedings Joint Residuals and Biosolids Management Conference*; Baltimore, Md., February 19–22, 2003; Water Environment Federation: Alexandria, Va., 2003.

Brandt, R.C.; Elliott, H.A.; O'Connor, G.A. Water Extractable Phosphorus in Biosolids: Implications for Land-based Recycling. *Water Environ. Res.* **2004**, *76*, 121–129.

Breeuwsma, A.; Reijerink, J.G.A.; Schoumans, O.F. Impact of Manure on Accumulation and Leaching of Phosphate in Areas of Intensive Livestock Farming; In *Animal Waste and the Land–Water Interface;* Steele, K.; Ed.; Lewis Publishers, CRC: New York, 1995; pp 238–251.

Bremner, J.M. Nitrogen-Total. In *Methods of Soil Analysis. Part 3*; Sparks, D.L., Ed.; SSSA Book Series no. 5, Soil Science Society of America and American Society of Agronomy: Madison, Wisc., 1996; pp 1085–1121.

Carpenter, S.R.; Caraco, N.F.;Correl, D.L.; Howarth, R.W.; Sharpley, A.N.;. Smith, V.H. Nonpoint Pollution of Surface Waters with Phosphorus and Nitrogen. *Ecol. Applic*. 1998, 8, 559–568.

Chardon, W.J.; Menon, R.G.; Chien, S.H. Iron Oxide Impregnated Filter Paper (Pi test): A Review of Its Development and Methodological Research. *Nutr. Cycl. Agroecosys*. **1996**, *46*, 41–51.

Clesceri, L.S.; Greenberg, A.E.; Eaton,A.D,; Eds. *Standard Methods for the Examination of Water and Wastewater;* 19th Edition; American Public Health Assoc., Washington, D.C., 1998.

Coale, F.J.; Sims, J.T. Leytem, A.B. Accelerated Deployment of an Agricultural Nutrient Management Tool: The Maryland Phosphorus Site Index. *J. Environ. Qual.* **2002**, *31*, 1471–1476.

Czapar, G.F.; Simmons, F.W. Water Quality, In *Illinois Agronomy Handbook*, Picklesimer, P., Ed.; University of Illinois, Urbana, Ill., 2002.

Daigger, G.T. Tools for Future Success: Emerging Trends that are Changing the Nature of Water Quality Management. *Water Environ. Tech.* 2003, 15(12), 38–45.

Dayton, E.A.; Basta, N.T.; Jakober, C.A.; Hattey, J.A. Using Water Treatment Residuals to Reduce Phosphorus in Agricultural Runoff. *J. Am. Water Works Assoc.* **2003**, *95*, 151–158.

Dayton, E.A.; Basta, N.T. Use of Drinking Water Treatment Residuals as a Potential Best Management Practice to Reduce Phosphorus Risk Index Scores. *J. Environ. Qual.* **2005,** *34*, 2112–2117.

De Haan, S. Sewage Sludge as a Phosphate Fertilizer. *Phosphorus Agric*. **1980**, *34*, 33–41.

Ebeling, A.M.; Cooperband, L.R.; Bundy, L.G. Phosphorus Availability to Wheat from Manures, Biosolids, and an Inorganic Fertilizer. *Commun. Soil Sci. Plt. Anal.* **2003**, *34*, 1347–1365.

Edwards, D.R.; Daniel, T.C. Runoff Quality Impacts of Swine Manure Applied to Fescue Plots. *Trans. ASAE* **1993**, *36*, 81–86.

Elliott, H.A.; O'Connor, G.A.; Lu, P.; Brinton, S. Influence of Water Treatment Residual on Phosphorus Solubility and Leaching. *J. Environ. Qual.* **2002**, *31*, 1362–1369.

Elliott, H.A.; Brandt, R.C.; O'Connor, G.A. Runoff Phosphorus Losses from Surface-applied Biosolids. *J. Environ. Qual.* **2005a**, *34*, 1632–1639.

Elliott, H.A.; Potter, J.M.; Kang, J.-H.; Brandt, R.C.; O'Connor, G.A. Neutral Ammonium Citrate Extraction of Biosolids Phosphorus. *Comm. Soil Sci. Plt. Anal*. **2005b**, *36*, 2447–2459.

Elliott, H.A.; Brandt, R.C.; Kleinman, P.J.A.; Sharpley, A.N.; Beegle, D.B. Estimating Source Coefficients for Phosphorus Site Indices. *J. Environ. Qual.* **2006**, *35*.

Epstein, E. *Land Application of Sewage Sludge and Biosolids*. Lewis Publishers: Boca Raton, Fla., 2003.

Gburek, W.J.; Sharpley, A.N.; Heathwaite, L.; Folmer, G.J. Phosphorus Management at the Watershed Scale: A Modification of the Phosphorus Index. *J. Environ. Qual.* **2000**, *29*, 130–144.

Hanlon, E.A.; Gonzalez, J.S.; Bartos, J.M. *Mehlich-1 Extractable P, Ca, Mg, Mn, Cu, and Zn. IFAS Extension Soil Testing Laboratory (ESTIL) and Analytical Research Laboratory (ARL) Chemical Procedures and Training Manual.* University of Florida/IFAS: Gainesville, Fla., 1997; pp 18–20.

Hesketh, N.; Brookes, P.C. Development of an Indicator for Risk of Phosphorus Leaching. *J. Environ. Qual.* **2000**, *29*, 105–110.

Higgs, B.; Johnston, A.E.; Salter, J.L.; Dawson, C.J. Some Aspects of Achieving Sustainable Phosphorus Use in Agriculture. *J. Environ. Qual.* **2000,** *29,* 80–87.

Jokinen, R. Effect of Phosphorus Precipitation Chemicals on Characteristics and Agricultural Value of Municipal Sewage Sludges. *Acta Agric. Scan.* **1990**, *40*, 141–147.

Kane, P.F., Ed. *Official Methods of Analysis of AOAC International, Chapter 2: Fertilizers*, Association of Official Analytical Chemists International: Arlington, Va., 1995.

Kleinman, P.; Sharpley, A.; Bryant, R.; Reid, S. Phosphorus and Agriculture IV: The National Phosphorus Project. *What's Cropping Up?* **2001**, *10*, 2–3.

Kleinman, P.J.A.; Bryant, R.B.; Reid, W.S.; Sharpley, A.N.; Pimentel, D. Using Soil Phosphorus Behavior to Identify Environmental Thresholds. *Soil Sci.* **2000**, *165*, 943–950.

Kleinman, P.J.A.; Sharpley, A.N.; Moyer, B.G.; Elwinger, G.F. Effect of Mineral and Manure Phosphorus Sources on Runoff Phosphorus. *J. Environ. Qual*. **2002a**, *31*, 2026–2033.

Kleinman, P.J.A.; Sharpley, A.N.; Wolf, A.M.; Beegle, D.B.; Moore, P.A. Measuring Water Extractable Phosphorus in Manure. *Soil Sci. Soc. Am. J.* **2002b**. *66*, 2009–2015.

Kleinman, P.J.A.; Sharpley, A.N. Effect of Broadcast Manure on Runoff Phosphorus Concentrations over Successive Rainfall Events. *J. Environ. Qual*. **2003**, *32*, 1072–1081.

Kleinman, P.J.A.; Sharpley, A.N.; Veith, T.L.; Maguire, R.O.; Vadas, P.A. Evaluations of Phosphorus Transport in Surface Runoff from Packed Soil Boxes. *J. Environ. Qual.* **2004**, *33*, 1413–1423.

Kleinman, P.J.A.; Wolf, A.M.; Sharpley, A.N.; Beegle, D.B.; Saporito, L.S. Survey of Water Extractable Phosphorus in Livestock Manures. *Soil Sci. Soc. Am. J.* **2005**, *69,* 701–708.

Kleinman, P.J.A. Personal communication, 2006.

Kyle, M.A.; McClintock, S.A. The Availability of Phosphorus in Municipal Wastewater Sludge as a Function of the Phosphorus Removal Process and the Sludge Treatment Method. *Water Environ. Res*. **1995**, *67*, 282–299.

Lemunyon, J.L.; Gilbert, R.G. The Concept and Need for a Phosphorus Assessment Tool. *J. Prod. Agric.* **1993**, *6*, 483–496.

Leytem, A.B.; Sims, J.T.; Coale, F.J. Determination of Phosphorus Source Coefficients for Organic Phosphorus Sources: Laboratory Studies. *J. Environ. Qual.* **2004**, *33*, 380–388.

Logan, J.A. Phosphorus runoff from biosolids-amended pastures during simulated runoff. M.S. Thesis, Pennsylvania State University: University Park, Pa., 2004.

Loosdrecht, M.V. Role of Biological Processes in Phosphate Recovery. *Proceedings of the International Conference on Phosphorus Recovery from Sewage and Animal Wastes.* May 6–7, 1998. University of Warwick: Conventry, U.K., 1998.

Maguire, R.O.; Sims, J.T. Measuring Agronomic and Environmental Soil Phosphorus Saturation and Predicting Phosphorus Leaching with Mehlich 3. *Soil Sci. Soc. Am. J.* **2002a**, *66*, 2033–2039.

Maguire, R.O.; Sims, J.T. Urban Waste and Sludges as Fertilizers. In *Encyclopedia of Soil Science*; Lal, R., Ed.; Marcel Dekker: New York, 2002b; pp 569–571.

Maguire, R.O.; Sims, J.T.; Coale, F.J. Phosphorus Solubility in Biosolids-amended Farm Soils in the Mid-Atlantic Region of the USA. *J. Environ. Qual.* **2000**, *29*, 1225–1233.

Maguire, R.O.; Sims, J.T.; Dentel, S.K.; Coale, F.J.; Mah, J.T. Relationships between Biosolids Treatment Process and Soil Phosphorus Availability. *J. Environ. Qual.* **2001,** *30*, 1023–1033.

Maine Department of Environmental Protection. Agricultural Utilization of Residuals, In *Maine Solid Waste Management Regulations*, MEDEP: Augusta, Maine, 1999; Chap. 419.

Martz, E., Ed. *The Agronomy Guide.* College of Agricultural Sciences, Pennsylvania State University: University Park, Pa., 2002.

McDowell, R.W.; Sharpley, A.N. Approximating Phosphorus Release from Soils to Surface Runoff and Subsurface Drainage. *J. Environ. Qual.* **2001**, *30*, 508–520.

McDowell, R.W.; Sharpley, A.N. The Effects of Soil Carbon on Phosphorus and Sediment Loss from Soil Trays by Overland Flow. *J. Environ. Qual.* **2003**, *32*, 207–214.

McKeague, J.A.; Brydon, J.E.; Miles, N.M. Differentiation of Forms of Extractable Iron and Aluminum in Soils. *Soil Sci. Soc. Am. Proc.* **1971**, *35*, 33–38.

Mehlich, A. Mehlich-3 Soil Test Extractant: A Modification of Mehlich-2 Extractant. *Commun. Soil Sci. Plant Anal.* **1984**, *15,* 1409–1416.

Moore, P.A.; Daniel, T.C; Edwards, D.R. Reducing Phosphorus Runoff and Improving Poultry Production with Alum. *Poultry Sci.* **1999**, *78*, 692–698.

Morris, R.; Smith, S.R.; Bellet-Travers, D.M.; Bell, J.N.B. Reproducibility of the Nitrogen Response and Residual Fertilizer Value of Conventional and Enhanced-treated Biosolids. *Proceedings, Joint Chartered Institution of Water and Environmental Management and Aqua. Environ. Technology Transfer, 8th European Biosolids and Organic Residuals Conference,* November 23–26, 2003; Wakefield, UK, 2003.

Murphy, J.; Riley, J.P. A Modified Single Solution Method for the Determination of Phosphate in Natural Water. *Anal. Chim. Acta* **1962**, *27*, 31–36.

Nair, V.D.; Portier, K.M.; Graetz, D.A.; Walker, M.L. An Environmental Threshold for Degree of Phosphorus Saturation in Sandy Soils. *J. Environ. Qual.* **2004**, *33*, 107–113.

National Phosphorus Research Project. National Research Project for Simulated Rainfall – Surface runoff studies. Southern Ext. Res. Activity Information Exchange Group 17, 2001. Available at: http://www.soil.ncsu.edu/sera17/publications/National_P/National_P_Project.htm.

O'Connor, G.A.; Sarkar, D.; Graetz, D.A.; Elliott, H.A. *Characterizing Forms, Solubilities, Bioavailabilities, and Mineralization Rates of Phosphorus in Biosolids, Commercial Fertilizers, and Manures*, Report 99-PUM-2. Water Environment Research Foundation: Alexandria, Va., 2002.

O'Connor, G.A.; Sarkar, D.; Brinton, S.R.; Elliott, H.A.; Martin, F.G. Phytoavailability of Biosolids-Phosphorus. *J. Environ. Qual.* **2004**, *33*, 703–712.

Ontario Ministry of Environment and Energy (OMEE) and Ontario Ministry of Agriculture, Food and Rural Affairs (OMAFRA). *Guidelines for the Utilization of Biosolids and Other Wastes on Agricultural Land.* OMME and OMAFRA: Toronto, Ontario, 1996.

Pastene, A.J. Factors Affecting the crop availability of phosphorus in sewage sludge amended soils. M.S. Thesis, University of Wisconsin, Madison, 1981.

Penn, C.J.; Sims, J.T. Phosphorus Forms in Biosolids-amended Soils and Losses in Runoff: Effects of Wastewater Treatment Process. *J. Environ. Qual.* **2002**, *31*, 1349–1361.

Pierson, S.T.; Cabrera, M.L.; Evanylo, G.K.; Kuykendall, H.A.; Hoveland, C.S.; McCann, M.A.; West, L.T. Phosphorus and Ammonium Concentrations in Surface Runoff from Grasslands Fertilized with Broiler Litter. *J. Environ. Qual.* **2001**, *30*, 1784–1789.

Plank, C.O. Plant Analysis Reference Procedures for the Southern Region of the United States. *Southern Coop. Ser. Bull. 1992*, 368.

Pote, D.H.; Daniel, T.C.; Nichols, D.J.; Sharpley, A.N.; Moore, Jr., P.A.; Miller, D.M.; Edwards, D.R. Relationship between Phosphorus Levels in Three Ultisols and Phosphorus Concentrations in Runoff. *J. Environ. Qual.* **1999**, *28*, 170–175.

Pote, D.H.; Reed, B.A.; Daniel, T.C.; Nichols, D.J.; Moore, Jr., P.A.; Edwards, D.R.; Formica, S. Water-Quality Effects of Infiltration Rate and Manure Application Rate for Soils Receiving Swine Manure. *J. Soil Water Cons.* **2001**, *56*, 32–37.

Richards, B.K.; Peverly, J.H.; Steenhuis, T.S.; Liebowitz, B.N. Effect of Processing Mode on Trace Elements in Dewatered Sludge Products. *J. Environ. Qual.* **1997** *26*, 782–788.

Rostagno, C.M.; Sosebee, R.E. Biosolids Application in the Chihuahuan Desert: Effects on Runoff Water Quality. *J. Environ. Qual.* **2001**, *30*, 160–170.

SAS Institute. *SAS/STAT User's Guide;* version 6; SAS Institute Inc.: Cary, NC, 1989.

Sarkar, D.; O'Connor, G.A. Plant and Soil Responses to Biosolids-Phosphorus in Two Florida Soils with High Phosphorus Content. *Commun. Soil Sci. Plt. Anal.* **2005**, *35*, 1569–1589.

Sharpley, A.N. An Innovative Approach to Estimate Bioavailable Phosphorus in Agricultural Runoff Using Iron Oxide-Impregnated Paper. *J. Environ. Qual.* **1993a**, *22*, 597–601.

Sharpley, A.N. Assessing Phosphorus Bioavailability in Agricultural Soils and Runoff. *Fertil. Res.* **1993b**, *36*, 259–272.

Sharpley, A.N. Rainfall Frequency and Nitrogen and Phosphorus in Runoff from Soil Amended with Poultry Litter. *J. Environ. Qual.* **1997**, *26*, 1127–1132.

Sharpley, A.N.; Moyer, B. Phosphorus Forms in Manure and Compost and their Release during Simulated Rainfall. *J. Environ. Qual.* **2000**, *29,* 1462–1469.

Sharpley, A.N.; McDowell, R.W.; Weld, J.L.; Kleinman, P.J.A. Assessing Site Vulnerability to Phosphorus Loss in an Agricultural Watershed. *J. Environ. Qual.* **2001,** *30,* 2026–2036.

Sharpley, A.N.; Weld, J.L.; Beegle, D.B.; Kleinman, P.J.A.; Gburek, W.J.; Moore, Jr., P.A.; Mullins, G. Development of Phosphorus Indices for Nutrient Management Planning Strategies in the United States. *J. Soil Water Conserv.* **2003,** *58*, 137–152.

Shober, A.L.; Sims, J.T. Phosphorus Restrictions for Land Application of Biosolids: Current Status and Future Trends. *J. Environ. Qual.* **2003,** *32,* 1955–1964.

Shreve, B.R.; Moore, P.A.; Miller, D.M.; Daniel, T.C.; Edwards, D.R. Long-term Phosphorus Solubility in Soils Receiving Poultry Litter Treated with Aluminum, Calcium, and Iron Amendments. *Commun. Soil Sci. Plt. Anal.* **1996,** *27*, 2493–2510.

Silveira, M.L.; Miyittah, M.K.; O'Connor, G.A. Phosphorus Release from a Manure-impacted Spodosol: Effects of a Water Treatment Residual. *J. Environ. Qual.* **2006,** *35*, 529–541.

Sims, J.T.; Coale, F.; Evanylo, G.; White, J.; Leytem, A. Field and On-Farm Evaluation of the Effects of Biosolids on Phosphorus in Soils and Runoff; Final Project Report. Metropolitan Washington Council of Governments. Washington, D.C., 2003.

Smith, S. Personal communication, June 2003.

Smith, S.R.; Triner, N.G.; Knight, J.J. Phosphorus Release and Fertilizer Value of Enhanced-treated and Nutrient Removal Biosolids. *Water Environ. Mgmt.* **2002,** *16*, 127–134.

Smith, S.R.; Morris, R.; Bellet-Travers, D.M.; Bell J.N.B. Managing the Phosphate Supply from Sewage Sludge Amended Agricultural Soil. *Proceedings, Joint Chartered Institution of Water and Environmental Management and Aqua. Environ. Technology Transfer, 8th European Biosolids and Organic Residuals Conference,* November 23–26, 2003; Wakefield, U.K., 2003

Soon, Y.K.; Bates; T.E. Extractability and Solubility of Phosphate in Soils Amended with Chemically Treated Sewage Sludges. *Soil Sci.* **1982,** *134,* 89–96.

Stehouwer, R.C.; Wolf, A.M.; Doty W.T. Chemical Monitoring of Sewage Sludge in Pennsylvania: Variability and Application Uncertainty. *J. Environ. Qual.* **2000**, *29,* 1686–1695.

Tchobanoglous, G.; Burton, F.L.; Stense, H.D.; Metcalf & Eddy, Inc. *Wastewater Engineering: Treatment and Reuse;* 4th Edition; McGraw Hill: Boston, Mass., 2003.

U.S. Department of Agriculture and U.S. Environmental Protection Agency (U.S. EPA). *Unified National Strategy for Animal Feeding Operations;* March 9, 1999; U.S. Government Printing Office: Washington, D.C., 1999.

U.S. EPA. *Process Design Manual. Land Application of Sewage Sludge and Municipal Septage;* EPA /625/R-95/001; Office of Research and Development: Cincinnati, Ohio, 1995.

U.S. EPA. *Test Methods for Evaluating Solid Waste, Physical/Chemical Methods;* U.S. EPA-65 FR 70678; Draft Update IVB SW-846; U. S. Governmental Printing Office: Washington, D.C., 2000.

Vadas, P.A.; Kleinman, P.J.A.; Sharpley, A.N.; Turner, B.L. Relating Soil Phosphorus to Dissolved Phosphorus in Runoff: A Single Extraction Coefficient for Water Quality Modeling. *J. Environ. Qual.* **2005,** *34,* 572–580.

Van der Zee, S.E.A.T.M.; L.G.J. Fokkink; van Riemsdjil, W.H. A New Technique for Assessment of Reversibly Adsorbed Phosphate. *Soil Sci. Soc. Am. J.* **1987,** *51,* 599–604.

Weld, J.L.; Beegle, D.B.; Sims, J.T.; Lemunyon, J.L.; Sharpley, A.N.; Gburek, W.J.; Coale, F.C.; Leytem, A.B. Comparison of State Approaches to the Development of Phosphorus Indices. In *2000 Annual Meetings Abstracts*; American Society of Agronomy, Crop Science Society of America, and Soil Science Society of America: Madison, Wisc., 2000; p 329.

Weld, J.L.; Beegle, D.B.; Gbruek, W.L.; Kleinman, P.J.A.; Sharpley, A.N. *The Pennsylvania Phosphorus Index:* Version 1; Pennsylvania State University: University Park, Pa., 2003. Available at: http://panutrientmgmt.cas.psu.edu/pdf/phosphorus_index_factsheet.pdf.

Welling, N., personal communication, 2005.

Withers, P.J.; Clay, S.D.; Breeze, V.G. Phosphorus Transfer in Runoff Following Application of Fertilizer, Manure, and Sewage sludge. *J. Environ. Qual.* **2001**, *30*, 180–188.

Withers, P.J.A.; Bailey, G.A. Sediment and Phosphorus Transfer in Overland Flow from a Maize Field Receiving Manure. *Soil Use Mgmt.* **2003**, *19*, 28–35.

Wolf, A.M.; Sharpley, A.N.; Beegle, D.B.; Saporito, L.S. Development of a Water-Extractable Phosphorus Test for Manure: An Interlaboratory Study. *Soil Sci. Soc. Am. J.* **2005,** *69,* 695–700.

WASTEWATER UTILITY

Alabama
Montgomery Water Works & Sanitary Sewer Board

Alaska
Anchorage Water & Wastewater Utility

Arizona
Glendale, City of, Utilities Department
Mesa, City of
Peoria, City of
Phoenix Water Services Department
Pima County Wastewater Management
Safford, City of

Arkansas
Little Rock Wastewater Utility

California
Calaveras County Water District
Central Contra Costa Sanitary District
Corona, City of
Crestline Sanitation District
Delta Diablo Sanitation District
Dublin San Ramon Services District
East Bay Dischargers Authority
East Bay Municipal Utility District
Eastern Municipal Water District
El Dorado Irrigation District
Fairfield-Suisun Sewer District
Fresno Department of Public Utilities
Inland Empire Utilities Agency
Irvine Ranch Water District
Las Virgenes Municipal Water District
Livermore, City of
Los Angeles, City of
Los Angeles County, Sanitation Districts of
Napa Sanitation District
Orange County Sanitation District
Palo Alto, City of
Riverside, City of
Sacramento Regional County Sanitation District
San Diego Metropolitan Wastewater Department, City of
San Francisco, City & County of
San Jose, City of
Santa Barbara, City of
Santa Cruz, City of
Santa Rosa, City of
South Bayside System Authority
South Coast Water District
South Orange County Wastewater Authority
Stege Sanitary District
Sunnyvale, City of
Union Sanitary District
West Valley Sanitation District

Colorado
Aurora, City of
Boulder, City of
Greeley, City of
Littleton/Englewood Water Pollution Control Plant
Metro Wastewater Reclamation District, Denver

Connecticut
Greater New Haven WPCA

District of Columbia
District of Columbia Water & Sewer Authority

Florida
Broward, County of
Fort Lauderdale, City of
Miami-Dade Water & Sewer Authority
Orange County Utilities Department
Reedy Creek Improvement District
Seminole County Environmental Services
St. Petersburg, City of
Stuart Public Utilities
Tallahassee, City of
Tampa, City of
Toho Water Authority
West Palm Beach, City of

Georgia
Atlanta Department of Watershed Management
Augusta, City of
Clayton County Water Authority
Cobb County Water System
Columbus Water Works
Fulton County
Gwinnett County Department of Public Utilities
Savannah, City of

Hawaii
Honolulu, City & County of

Idaho
Boise, City of

Illinois
American Bottoms Wastewater Treatment Plant
Greater Peoria Sanitary District
Kankakee River Metropolitan Agency
Metropolitan Water Reclamation District of Greater Chicago
Wheaton Sanitary District

Iowa
Ames, City of
Cedar Rapids Wastewater Facility
Des Moines, City of
Iowa City

Kansas
Johnson County Unified Wastewater Districts
Unified Government of Wyandotte County/ Kansas City, City of

Kentucky
Louisville & Jefferson County Metropolitan Sewer District

Louisiana
Sewerage & Water Board of New Orleans

Maine
Bangor, City of
Portland Water District

Maryland
Anne Arundel County Bureau of Utility Operations
Howard County Department of Public Works
Washington Suburban Sanitary Commission

Massachusetts
Boston Water & Sewer Commission
Upper Blackstone Water Pollution Abatement District

Michigan
Ann Arbor, City of
Detroit, City of
Holland Board of Public Works
Saginaw, City of
Wayne County Department of Environment
Wyoming, City of

Minnesota
Rochester, City of
Western Lake Superior Sanitary District

Missouri
Independence, City of
Kansas City Missouri Water Services Department
Little Blue Valley Sewer District
Metropolitan St. Louis Sewer District

Nebraska
Lincoln Wastewater System

Nevada
Henderson, City of
Reno, City of

New Jersey
Bergen County Utilities Authority
Ocean, County of
Passaic Valley Sewerage Commissioners

New York
New York City Department of Environmental Protection

North Carolina
Charlotte/Mecklenburg Utilities
Durham, City of
Metropolitan Sewerage District of Buncombe County
Orange Water & Sewer Authority

Ohio
Akron, City of
Butler County Department of Environmental Services
Columbus, City of
Metropolitan Sewer District of Greater Cincinnati
Northeast Ohio Regional Sewer District
Summit, County of

Oklahoma
Oklahoma City Water & Wastewater Utility Department
Tulsa, City of

Oregon
Clean Water Services
Eugene/Springfield Water Pollution Control
Gresham, City of
Water Environment Services

Pennsylvania
Philadelphia, City of
University Area Joint Authority

South Carolina
Charleston Commissioners of Public Works
Mount Pleasant Waterworks & Sewer Commission
Spartanburg Sanitary Sewer District

Tennessee
Cleveland, City of
Knoxville Utilities Board
Murfreesboro Water & Sewer Department
Nashville Metro Water Services

Texas
Austin, City of
Dallas Water Utilities
Denton, City of
El Paso Water Utilities
Fort Worth, City of
Gulf Coast Waste Disposal Authority
Houston, City of
San Antonio Water System
Trinity River Authority

Utah
Salt Lake City Corporation

Virginia
Alexandria Sanitation Authority

WERF SUBSCRIBERS

Arlington, County of
Fairfax County
Hampton Roads Sanitation District
Henrico, County of
Hopewell Regional Wastewater Treatment Facility
Loudoun County Sanitation Authority
Lynchburg Regional WWTP
Prince William County Service Authority
Richmond, City of
Rivanna Water & Sewer Authority

Washington
Everett, City of
King County Department of Natural Resources
Seattle Public Utilities
Sunnyside, Port of
Yakima, City of

Wisconsin
Green Bay Metro Sewerage District
Kenosha Water Utility
Madison Metropolitan Sewerage District
Milwaukee Metropolitan Sewerage District
Racine, City of
Sheboygan Regional Wastewater Treatment
Wausau Water Works

Australia
South Australian Water Corporation
Sydney Water Corporation
Water Corporation of Western Australia

Canada
Greater Vancouver Regional District
Regina, City of, Saskatchewan
Toronto, City of, Ontario
Winnipeg, City of, Manitoba

New Zealand
Watercare Services Limited

United Kingdom
Yorkshire Water Services Limited

STORMWATER UTILITY

California
Los Angeles, City of, Department of Public Works
Monterey, City of
Sacramento, County of
San Francisco, City & County of
Santa Rosa, City of
Sunnyvale, City of

Colorado
Aurora, City of
Boulder, City of

Georgia
Griffin, City of

Iowa
Cedar Rapids Wastewater Facility
Des Moines, City of

Kansas
Overland Park, City of

Maine
Portland Water District

Minnesota
Western Lake Superior Sanitary District

North Carolina
Charlotte, City of, Stormwater Services

Pennsylvania
Philadelphia, City of

Tennessee
Chattanooga Stormwater Management

Texas
Harris County Flood Control District, Texas

Washington
Bellevue Utilities Department
Seattle Public Utilities

STATE

Arkansas Department of Environmental Quality
Connecticut Department of Environmental Protection
Fresno Metropolitan Flood Control District
Kansas Department of Health & Environment
Kentucky Department of Environmental Protection
Ohio River Valley Sanitation Commission
Urban Drainage & Flood Control District, CO

CORPORATE

ADS Environmental Services
Alan Plummer & Associates
Alden Research Laboratory Inc.
Alpine Technology Inc.
Aqua-Aerobic Systems Inc.
Aquateam–Norwegian Water Technology Centre A/S
ARCADIS
Associated Engineering
Black & Veatch
Boyle Engineering Corporation
Brown & Caldwell
Burgess & Niple, Ltd.
Burns & McDonnell
CABE Associates Inc.
The Cadmus Group
Camp Dresser & McKee Inc.
Carollo Engineers Inc.
Carpenter Environmental Associates Inc.
CDS Technologies Inc.
CET Engineering Services
Chemtrac Systems Inc.
CH2M HILL
Conestoga-Rovers & Associates
CONTECH Stormwater Solutions
D&B/Guarino Engineers, LLC
Damon S. Williams Associates, LLC
Earth Tech Inc.
Ecovation
EMA Inc.
Environ/The ADVENT Group, Inc.
Fay, Spofford, & Thorndike Inc.
Freese & Nichols Inc.
ftn Associates Inc.
Fuss & O'Neill Inc.
Gannett Fleming Inc.
Geosyntec Consultants
GHD
Golder Associates Ltd.
Greeley and Hansen LLC
Hazen & Sawyer, P.C.
HDR Engineering Inc.
HNTB Corporation
Hydromantis Inc.
HydroQual Inc.
Infilco Degremont Inc.
Jacobson Helgoth Consultants Inc.
Jason Consultants LLC Inc.
Jordan, Jones, & Goulding Inc.
KCI Technologies Inc.
Kelly & Weaver, P.C.
Kennedy/Jenks Consultants
KMK Consultants
Komline Sanderson Engineering Corporation
Limno-Tech Inc.
Lombardo Associates Inc.
The Low Impact Development Center Inc.
Malcolm Pirnie Inc.
Material Matters
McKim & Creed
MEC Analytical Systems, Inc.
Metcalf & Eddy Inc.
Monteco Corporation
MPR Engineering Corporation, Inc
MWH
O'Brien & Gere Engineers Inc.
Odor & Corrosion Technology Consultants Inc.
Oscar Larson & Associates
PA Government Services Inc.
Parametrix Inc.
Parsons
Post, Buckley, Schuh & Jernigan
RMC Water & Environment
R.M. Towill Corporation
Ross & Associates Ltd.
Rothberg, Tamburini & Windsor, Inc.
SAIC
Siemens Water Technologies
Stantec Consulting Inc.
Stearns & Wheler, LLC
Stone Environmental Inc.
Stratus Consulting Inc.
Synagro Technologies Inc.
Tetra Tech Inc.
Trojan Technologies Inc.
Trussell Technologies, Inc.
URS Corporation
Wade-Trim Inc.
Westin Engineering Inc.
Weston Solutions Inc.
Woodard & Curran
Zenon Environmental Inc.
Zoeller Pump Company

INDUSTRY

American Electric Power
American Water
ChevronTexaco Energy Research & Technology Company
The Coca-Cola Company
Dow Chemical Company
DuPont Company
Eastman Chemical Company
Eastman Kodak Company
Eli Lilly & Company
Merck & Company Inc.
Premier Chemicals LLC
Procter & Gamble Company
RWE Thames Water Plc
Severn Trent Services Inc.
Suez Environnment
United Water Services LLC

Note: List as of 9/1/06

WERF Board of Directors

WERF Research Council

IWA Publishing's authorised EU representative for General Product Safety Regulations is Diane D'Arras, 15 rue Duret, 75116 Paris, France, e-mail: safety@iwap.co.uk.

Printed and bound by CPI Group (UK) Ltd, Croydon, CR0 4YY
12/05/2026
02108874-0011